Nicholson's guides to the WATERWAYS

South west

ROBERT NICHOLSON PUBLICATIONS

COMMUNICA-EUROPA

Robert Nicholson Publications
24 Highbury Crescent,
London N5.

Editors: Paul Atterbury and Andrew Darwin

Printed and bound in Great Britain by Morrison & Gibb Ltd, London and Edinburgh

Robert Nicholson Publications would like to thank all those who made expert contributions towards the preparation of this book, especially the staff of BWB.

Further copies of this publication, and others in the series, can be obtained from your local bookshop or direct from the British Waterways Board, Willow Grange, Church rd, Watford WD1 3QA.

ISBN 0 900568 21 6

R–12/75–10

Front cover: A pair of narrowboats on the Worcester & Birmingham Canal. (Photo: Tom Sage)

Introduction

Canals today are enjoying a boom in popularity unrivalled since the halcyon days of the 1790's when 'canal mania' was at its peak. Then, canals were being built all over the country as efficient and profitable new transport routes; but nowadays, when trading on narrow canals is almost finished, the waterways of Britain have obtained a new lease of life. Large numbers of people are beginning to realise what a minority has known for a long time: that in a bustling, noisy, dangerous world, the canals provide a unique antidote to so much that annoys, bores and disturbs us. A 2,000 mile network woven across the country, the canals are a part of the English landscape and a part of our history. Quiet and dignified, the rural canal with its charming bridges, archaic locks and colourful boats has achieved a unity with, and enhancement of, the countryside in a way that nothing since the 18thC has quite equalled.

The contrast between rural and urban canals is dramatic, for it is in a town that the canal truly enters a world of its own. Slipping unnoticed through the town, the canal affords a rare and rewarding glimpse of the Industrial Revolution; and it is here that not only boatmen but walkers, industrial archaeologists and even fishermen find plenty of scope for their quiet enjoyment unspoilt by the bustle and noise of traffic. It is encouraging to see that more and more local councils are making a real effort to use and improve their canal as a valuable amenity.

There is interest for everyone in the canals: the engineering feats like aqueducts, tunnels and flights of locks (all of which amazed a world which had seen nothing like it since Roman times); the brightly decorated narrowboats which used to throng the waterways; the wealth of birds, animals and plants on the canal banks; the mellow, unpretentious architecture of canalside buildings like pubs, stables, lock cottages and warehouses; and the sheer beauty and quiet isolation that is a feature of so many English canals.

Contents

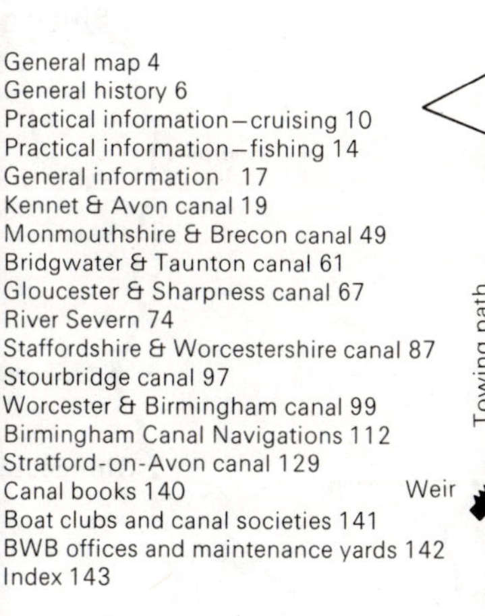

Symbols

N North axis

N North axis where map is cut

Tunnel

B Boatyard

28 8' 8" Lock, lock number and rise

Restaurant

Wine licence

Pub

63 Bridge and bridge number

R Refuse disposal

S Sewage disposal

W Water point

P Petrol

D Diesel

G Groceries

Winding hole: turning point for boats longer than the ordinary width of the canal

Towing path

Weir

Next to each map will be found a description of the countryside, plus places of interest and details regarding pubs, restaurants, boatyards etc.

The distance given under the name of the section is the approximate mileage on that particular page

The maps are 2 inches to 1 mile scale.

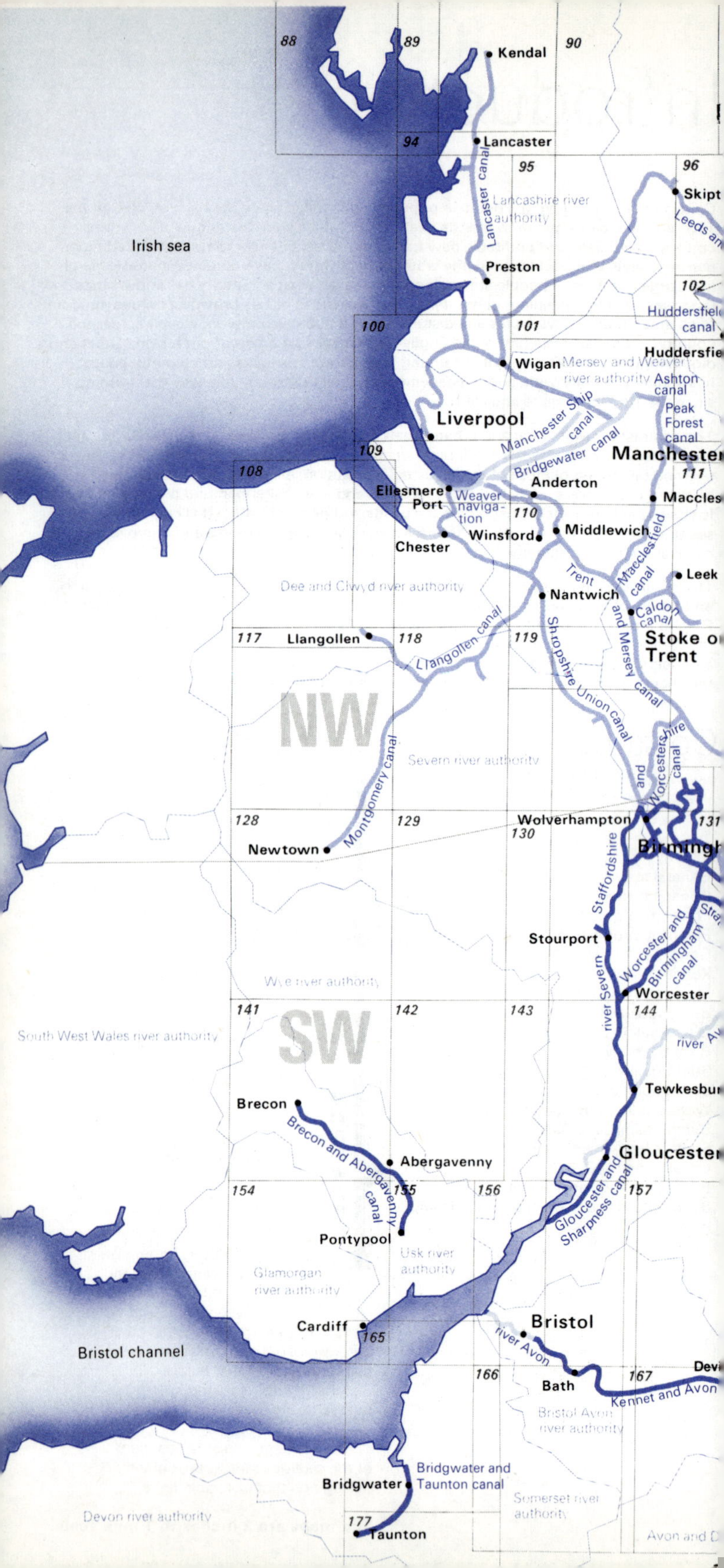

88
89
90
Kendal
94
Lancaster
95
96
Skipt
Lancaster canal
Lancashire river authority
Leeds an
Irish sea
Preston
102
Huddersfiel canal
100
101
Huddersfie
Wigan
Mersey and Weaver river authority
Ashton canal
Peak Forest canal
Liverpool
Manchester Ship canal
Bridgewater canal
Manchester
109
108
111
Ellesmere Port
Weaver navigation
Anderton
Maccles
110
Chester
Winsford
Middlewich
Macclesfield canal
Leek
Dee and Clwyd river authority
Trent and Mersey canal
Nantwich
Caldon canal
117
Llangollen
118
119
Stoke o Trent
Llangollen canal
Shropshire Union canal
NW
Staffordshire and Worcestershire canal
Severn river authority
Montgomery canal
128
129
Wolverhampton
131
130
Newtown
Birmingh
Staffordshire
Stourport
Worcester and Birmingham canal
Wye river authority
river Severn
Worcester
141
142
143
144
South West Wales river authority
SW
river Av
Tewkesbu
Brecon
Brecon and Abergavenny canal
Abergavenny
Gloucester
154
155
156
157
Gloucester and Sharpness canal
Pontypool
Usk river authority
Glamorgan river authority
Cardiff
165
Bristol
Bristol channel
river Avon
166
Bath
167
Devi
Kennet and Avon
Bristol Avon river authority
Bridgwater and Taunton canal
Bridgwater
Somerset river authority
Devon river authority
177
Taunton
Avon and D

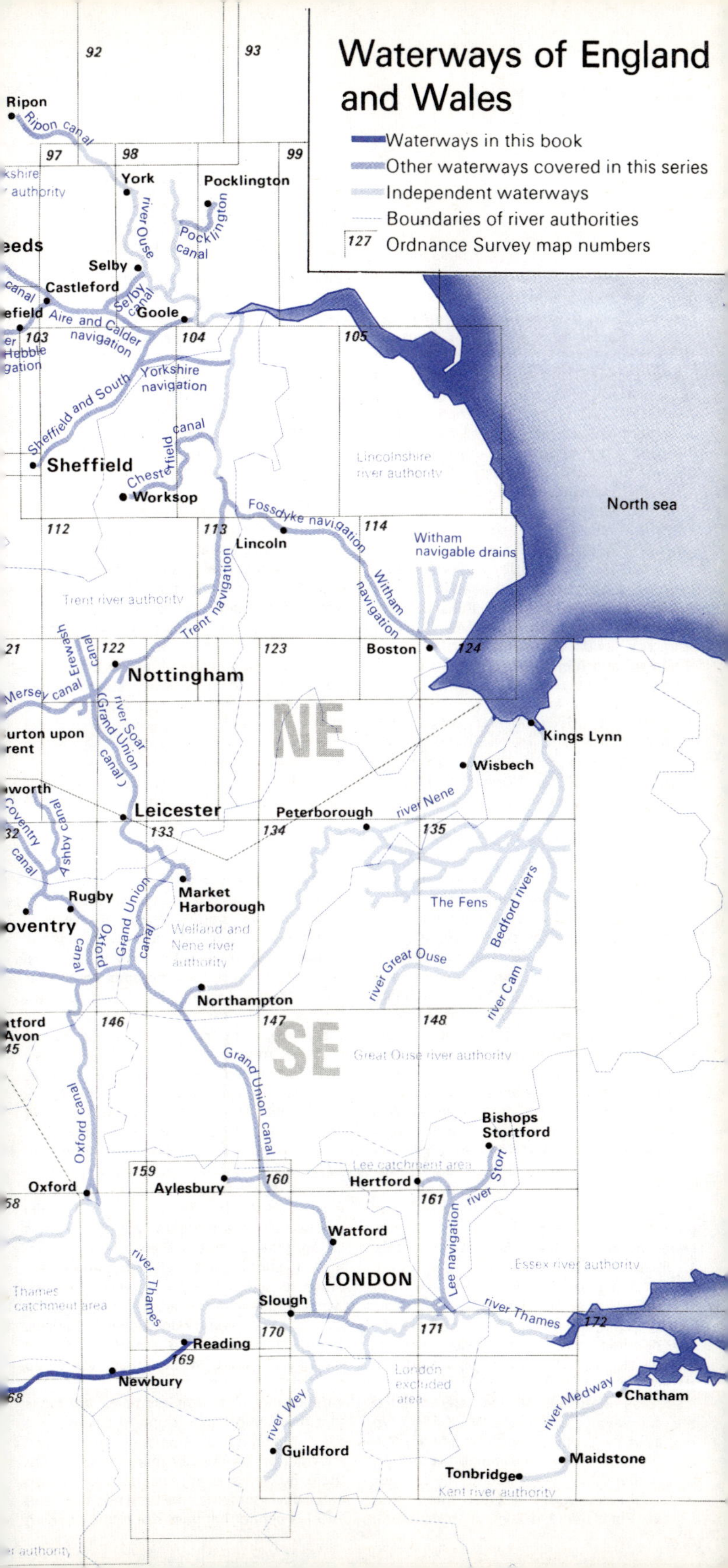

Waterways of England and Wales
Waterways in this book
Other waterways covered in this series
Independent waterways
Boundaries of river authorities
127 Ordnance Survey map numbers
Ripon
Ripon canal
York
Pocklington
Pocklington canal
river Ouse
Selby
Selby canal
Castleford
Goole
Aire and Calder navigation
Sheffield and South Yorkshire navigation
Sheffield
Chesterfield canal
Worksop
Lincolnshire river authority
North sea
Fossdyke navigation
Lincoln
Witham navigable drains
Witham navigation
Trent river authority
Trent navigation
Boston
Nottingham
Erewash canal
Mersey canal
river Soar (Grand Union canal)
NE
Kings Lynn
Wisbech
Leicester
Peterborough
river Nene
Coventry canal
Ashby canal
Rugby
Market Harborough
The Fens
Bedford rivers
Oxford canal
Grand Union canal
Welland and Nene river authority
river Great Ouse
river Cam
Northampton
SE
Great Ouse river authority
Grand Union canal
Oxford canal
Bishops Stortford
Lee catchment area
Oxford
Aylesbury
Hertford
river Stort
Watford
Lee navigation
Essex river authority
LONDON
Thames catchment area
river Thames
Slough
river Thames
Reading
Newbury
London excluded area
river Wey
river Medway
Chatham
Guildford
Maidstone
Tonbridge
Kent river authority

History

River navigations, that is rivers widened and deepened to take large boats, had existed in England since the middle ages; some can even be traced back to Roman times. In 1600 there were 700 miles of navigable river in England, and by 1760, the dawn of the canal age, this number had been increased to 1300. This extensive network had prompted many developments later used by the canal engineers, for example the lock system. But there were severe limitations; generally the routes were determined by the rivers and the features of the landscape and so were rarely direct. Also there were no east-west, or north-south connections.

Thus the demand for a direct inland waterway system increased steadily through the first half of the 18thC with the expansion of internal trade. Road improvements could not cope with this expansion, and so engineers and merchants turned to canals, used extensively on the continent.

One of the earliest pure canals, cut independently of existing rivers, was opened in 1745, at Newry in Northern Ireland, although some authorities consider the Fossdyke, cut by the Romans to link the rivers Trent and Witham, to be the first. However, the Newry is more important because it established the cardinal rule of all canals, the maintenance of an adequate water supply, a feature too often ignored by later engineers. The Newry canal established the principle of a long summit level, fed by a reservoir to keep the locks at either end well supplied. Ten years later, in England, the Duke of Bridgewater decided to build a canal to provide an adequate transport outlet for his coal mines at Worsley. He employed the self-taught James Brindley as his engineer, and John Gilbert as surveyor, and launched the canal age in England. The Bridgewater canal was opened in 1761. Its route, all on one level, was independent of all rivers; its scale of operations reflected the new power of engineering, and the foresight of its creators. Although there were no locks, the engineering problems were huge; an aqueduct was built at Barton over the river Irwell, preceded by an embankment 900 yards long; 15 miles of canal were built underground, so that boats could approach the coal face for loading—eventually there were 42 miles underground, including an inclined plane—the puddled clay method was used by Brindley to make the canal bed watertight. Perhaps most important of all, the canal was a success financially. Bridgewater invested the equivalent of £3 million of his own money in the project, and still made a profit.

Having shown that canals were both practical and financially sound, the Bridgewater aroused great interest throughout Britain. Plans were drawn up for a trunk canal, to link the four major rivers of England: the Thames, Severn, Mersey and Trent. This plan was eventually brought to fruition, but many years later than its sponsors imagined. Brindley was employed as engineer for the scheme, his reputation ensuring that he would always have more work than he could ever handle. The Trent and Mersey, and the Staffordshire and Worcestershire canals received the Royal Assent in 1766, and the canal age began in earnest.

Canals, like the railways later, were built entirely by hand. Gangs of itinerant workmen were gathered together, drawn by the comparatively high pay. Once formed, these armies of 'navigators'—hence 'navvies'—moved through the countryside as the canal was built, in many cases living off the land. All engineering problems had to be solved by manpower alone, aided by the horse and the occasional steam pump. Embankments, tunnels, aqueducts, all were built by these labouring armies kept under control only by the power of the section engineers and contractors.

The Staffordshire and Worcestershire canal opened in 1770. In its design Brindley determined the size of the standard Midlands canal, which of course had direct influence on the rest of the English system as it was built. He chose a narrow canal, with locks 72ft 7in by 7ft 6in, partly for reasons of economy, and partly because he realised that the problems of an adequate water supply were far greater than most canal sponsors realised. This standard, which was also adopted for the Trent and Mersey, prompted the development of a special vessel, the narrow boat with its 30-ton payload. Ironically this decision by Brindley in 1766 ensured the failure of the canals as a commercial venture 200 years later, for by the middle of this century a 30-ton payload could no longer be worked economically.

The Trent and Mersey was opened in 1777. 93 miles long, the canal included 5 tunnels, the one at Harecastle taking 11 years to build. In 1790 Oxford was finally reached and the junction with the Thames brought the four great rivers together. From the very start English canal companies were characterised by their intense rivalries; water supplies were jealously guarded, and constant wars were waged over toll prices. Many canals receiving the Royal Assent were never built, while others staggered towards conclusion, hampered by doubtful engineering, inaccurate estimates, and loans that they could never hope to pay off. Yet for a period canal mania gripped British speculators, as railway mania was to grip them 50 years later. The peak of British canal development came between 1791 and 1794, a period that gave rise to the opening

'Navvies' (navigators) at work, removing the stop lock at Braunston in the early 1930's.

of the major routes, the rise of the great canal engineers, Telford, Rennie and Jessop, and the greatest prosperity of those companies already operating. At this time the canal system had an effective monopoly over inland transport; the old trunk roads could not compete, coastal traffic was uncertain and hazardous, and the railways were still a future dream. This period also saw the greatest feats of engineering.

The turn of the century saw the opening of the last major cross country routes; the Pennines were crossed by the Leeds and Liverpool canal between 1770 and 1816 while the Kennet and Avon, opened in 1810, linked London and Bristol via the Thames. These two canals were built as broad navigations: already the realisation was dawning on canal operators that the limits imposed by the Brindley standard were too restricting, a suspicion that was to be brutally confirmed by the coming of the railways. The Kennet and Avon, along with its rival the Thames and Severn, also marks the introduction of fine architecture to canals. Up till now canal architecture had been functional, often impressive, but clearly conceived by engineers. As a result the Kennet and Avon has an architectural unity lacking in earlier canals—incidentally another reason to justify its preservation. The appearance of architectural quality was matched by another significant change; canals became straighter, their engineers choosing as direct a route as possible, arguing that greater construction costs would be outweighed by smoother, quicker operation, whereas the early canals had followed the landscape. The Oxford is the prime example of a contour canal, meandering across the midlands as though there were all the time in the world. It looks beautiful, its close marriage with the landscape makes it ideal as a pleasure waterway, but it was commercial folly.

The shortcomings of the early canals were exploited all too easily by the new railways. At first there was sharp competition by canals. Tolls were lowered, money was poured into route improvements; 14 miles of the Oxford's windings were cut out between 1829 and 1834; schemes were prepared to widen the narrow canals; the Harecastle tunnel was doubled in 1827, the new tunnel taking 3 years to build (as opposed to the 11 years of the old). But the race was lost from the start. The 19thC marks the rise of the railways and the decline of the canals. With the exception of the Manchester Ship Canal, the last major canal was the Birmingham and Liverpool Junction, opened in 1835. The system survived until this century, but the 1914–18 war brought the first closures, and through

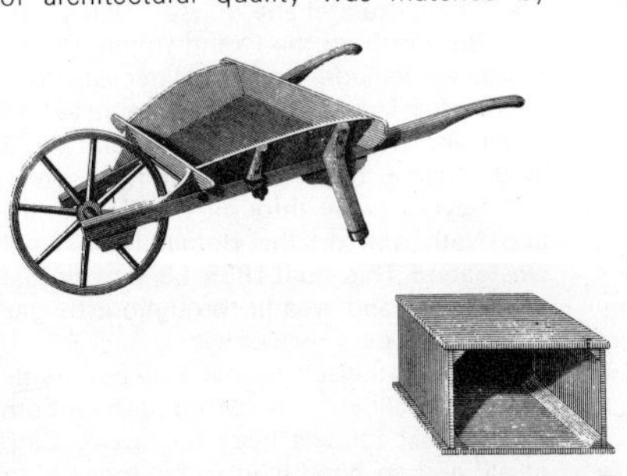

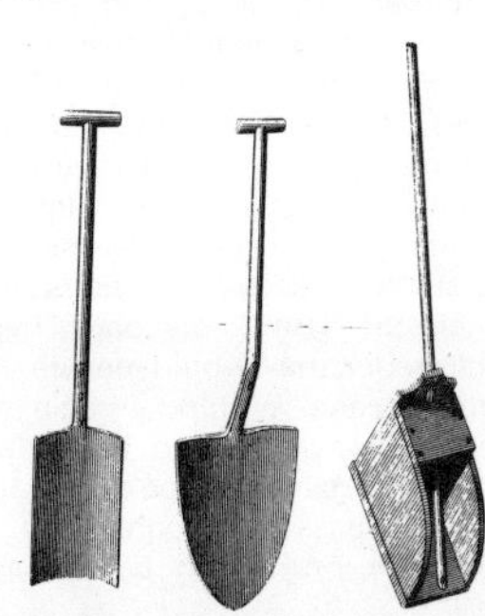

The rudimentary tools of the early 'navvies'. *Hugh McKnight.*

Worcester and Birmingham Canal.

LOWESMERE WHARF, WORCESTER.

Canal Company's Machine. 181

Mr.

Cheque, No.

Carriage, No.

Bought of

Gross Weight . . Tons. Cwt. Qrs.

Tare . . . .

Neat Coals, Cwt.

At per Ton £ s. d.

Weighing

Carriage

Clerk.

Clerk.

Printed by R. Jabet and Co. Herald Office, High-street, Birmingham.

Worcester and Birmingham Canal Company toll ticket, dated 1816. *Hugh McKnight.*

the 1930's the canal map adopted the shape it has today. Effective commercial carrying on narrow canals ceased in the early 1960's although a few companies managed to survive until recently. However, with the end of commercial operation, a new role was seen for the waterways, as a pleasure amenity, a 'linear national park 2000 miles long'.

Water supply has always been the cardinal element in both the running and the survival of any canal system. Locks need a constant supply of water—every boat passing through a wide lock on the Grand Union uses 96,000 gallons of water. Generally two methods of supply were used: direct feed by rivers and streams, and feed by reservoirs sited along the summit level. The first suffered greatly from silting, and meant that the canal was dependent on the level of water in the river; the regular floods from the river Soar that overtake the Grand Union's Leicester line show the dangers of this. The second was more reliable, but many engineers were short-sighted in their provision of an adequate summit level. The otherwise well planned Kennet and Avon always suffered from water shortage. Where shortages occurred, steam pumping engines were used to pump water taken down locks back up to the summit level. The Kennet and Avon was dependent upon pumped supplies, while the Birmingham Canal Navigations were fed by 6 reservoirs and 17 pumping engines. Some companies adopted side ponds alongside locks to save water, but this put the onus on the boatman and so had limited success. Likewise the stop locks still to be seen at junctions are a good example of 18thC company rivalry; an established canal would ensure that any proposed canal wishing to join it would have to lock *down* into the older canal, which thus gained a lock of water each time a boat passed through.

Where long flights or staircase locks existed there was always great wastage of water, and so throughout canal history alternative mechanical means of raising boats have been tried out. The inclined plane or the vertical lift were the favoured form. Both worked on the counterbalance principle, the weight of the descending boat helping to raise the ascending. The first inclined plane was built at Ketley in 1788, and they were a feature of the west country Bude and Chard canals. The most famous plane was built at Foxton, and operated from 1900-1910. Mechanical failure and excessive running costs ended the application of the inclined plane in England, although modern examples work very efficiently on the continent, notably in Belgium. The vertical lift was more unusual, although there were 8 on the Grand Western canal. The most famous, built at Anderton in 1875, is still in operation, and stands as a monument to the ingenuity shown in the attempts to overcome the problems of water shortage.

Engineering features are the greatest legacy of the canal age, and of these, tunnels are the most impressive. The longest tunnel is at Standedge, on the now derelict Huddersfield narrow canal. The tunnel runs for 5456 yards through the Pennines, at times 600ft below the surface. It is also on the highest summit level, 656ft above sea level. The longest tunnel still in use, 3056 yards, is at Blisworth on the Grand Union. Others of interest include the twin Harecastle tunnels on the Trent and Mersey, the first 2897 yards and now disused, the second 2926 yards, Sapperton which carried the Thames and Severn canal through the Cotswolds, and Netherton on the Birmingham Canal Navigation. This, built 1855-58, was the last in England, and was lit throughout by gas lights, and later by electricity.

The Netherton tunnel was built wide enough to allow for a towing path on both sides. Most tunnels have no towing path at all, and so boats had to be 'legged', or

Islington tunnel under construction. *Hugh McKnight.*

walked through.

The slowness and relative danger of legging in tunnels led to various attempts at mechanical propulsion. An endless rope pulled by a stationary steam engine at the tunnel mouth was tried out at Blisworth and Braunston between 1869 and 1871. Steam tugs were employed, an early application of mechanical power to canal boats, but their performance was greatly limited by lack of ventilation, not to mention the danger of suffocating the crew.

An electric tug was used at Harecastle from 1914 to 1954. The diesel engine made tunnel tug services much more practical, but diesel powered narrow boats soon put the tugs out of business; by the 1930's most tunnels had to be navigated by whatever means the boatman chose to use. Legging continued at Crick, Husbands Bosworth and Saddington until 1939.

Until the coming of the diesel boats, the horse reigned supreme as a source of canal power. The first canals had used gangs of men to bow-haul the boats, a left over from the river navigations where 50-80 men, or 12 horses, would pull a 200-ton barge. By 1800 the horse had taken over, and was used throughout the heyday of the canal system. In fact horse towage survived as long as large scale commercial operation. Generally one horse or one mule was used per boat, a system unmatched for cheapness and simplicity. The towing path was carried from one side of the canal to the other by turnover bridges, a common feature that reveals the total dominance of the horse. Attempts to introduce self-propelled canal boats date from 1793, although most early experiments concerned tugs towing dumb barges. Development was limited by the damage caused by wash, a problem that still applies today, and the first fleets of self propelled steam narrow boats were not in service until the last quarter of the 19thC. Fellows, Morton and Clayton, and the Leeds and Liverpool Carrying Co. ran large fleets of steam boats between 1880 and 1931, by which time most had been converted to diesel operation. With the coming of mechanical power the butty boat principle was developed; a powered narrow boat would tow a dumb 'butty' boat, thereby doubling the load without doubling the running costs. This system became standard until the virtual ending by the late 1960's of carrying on the narrow canals. Before the coming of railways, passenger services were run on the canals; packet boats, specially built narrow boats with passenger accommodation, ran express services, commanding the best horses and the unquestioned right of way over all other traffic. Although the railways killed this traffic, the last scheduled passenger service survived on the Gloucester and Berkeley canal until 1935.

The traditional narrow boat with its colourful decoration and meticulous interior has become a symbol of English canals. However this was in fact a late development. The shape of the narrow boat was determined by Brindley's original narrow canal specification, but until the late 19thC boats were unpainted, and carried all male crews. Wages were sufficient for the crews to maintain their families at home. The increase in railway competition brought a reduction in wages, and so bit by bit the crews were forced to take their families with them, becoming a kind of water gipsy. The confines of a narrow boat cabin presented the same problems as a gipsy caravan, and so the families found a similar answer. Their eternally wandering home achieved individuality by extravagant and colourful decoration, and the traditional narrow boat painting was born. The extensive symbolic vocabulary available to the painters produced a sign language that only these families could understand, and the canal world became far more enclosed, although outwardly it was more decorative. As the canals have turned from commerce to pleasure, so the traditions of the families have died out, and the families themselves have faded away. But their language survives, although its meaning has mostly vanished with them. This survival gives the canals their characteristic decorative qualities which make them so attractive to the pleasure boatman and to the casual visitor.

Practical information cruising

Licences
Pleasure craft using the Board's canals must be licensed, and those using the Board's rivers must be registered under the British Waterways Act 1971. Charges are based on the length of the boat, and a canal craft licence covers all the navigable waterways under the Board's control. Permits for permanent mooring on the canals are also issued by the Board. Apply in each case to:

Craft Licensing Office,
British Waterways Board,
Willow Grange,
Church Road,
Watford WD1 3QA.
(Watford 26422).

The Licensing Officer will supply you with a list of BWB navigations.

Getting afloat
There is no better way of finding out the joys of canals than by getting afloat. The best thing is to hire a boat for a week or a fortnight from one of the boatyards on the canals. (Each boatyard has an entry in the text, and most of them offer hire cruisers; brochures may be easily obtained from such boatyards). A list of boatyard addresses is available from the BWB at Watford.

General cruising
Most canals are saucer shaped in section and so are deepest in the middle. However very few narrow canals have more than 3-4ft of water and many have much less. Try to keep to the middle of the channel on such waterways except at bends, where the deepest water is on the *outside* of the bend. When you meet another boat, the rule of the road is to keep to the right; slow down, and aim to miss the approaching boat by a couple of yards; do not steer right over to the bank unless the channel is particularly narrow or badly overgrown, or you will most likely run aground. The deeper the draught of the boat, the more important it is to keep in the middle of the deep water, and so this must be considered when passing other boats. If you meet a loaded working boat, keep right out of the way. Working boats should always be given precedence, for their time is money. If you meet a boat being towed from the bank, pass it on the outside rather than intercept the towing line. When overtaking, keep the other boat on your starboard, or right, side.

Speed
There is a general speed limit of 4mph on most British Waterways Board canals. This is not just an arbitrary limit; there is no need to go any faster, and in many cases it is impossible to cruise even at this speed. Canals were not built for motor boats, and so the banks are easily damaged by excessive wash and turbulence. Erosion of the banks makes the canal more shallow, which in turn makes running aground a more frequent occurrence. So keep to the limits and try not to aggravate the situation. It is easy to see when a boat is creating excessive turbulence by looking at the wash. If in doubt, slow down.

Slow down also when passing moored craft, engineering works and anglers.

Slow down when approaching blind corners, narrow bridges and junctions.

When approaching floating rubbish, especially plastic bags, speed up. Then disengage the engine so the boat 'glides' over the flotsam without the propeller becoming snarled up.

Running aground
The effective end of commercial traffic on the narrow canals has meant a general reduction in standards of dredging. These canals are now shallower than ever, and contain more rubbish than ever. Running aground is a common event, but is rarely serious, as the canal bed is usually soft. If you run aground, try first of all to pull the boat off by reversing the engine. If this fails, use the boat hook as a lever against the bank or some solid object, in combination with a tow rope being pulled from the bank. Do not keep revving the engine in reverse if it is obviously having no effect; this will merely damage both your propeller and the canal bed by drawing water away from the boat. Another way is to get your crew to rock the boat from side to side while using the boat hook or tow rope. If all else fails, lighten your load; make all the crew leave the boat except the helmsman, and then it will often float off quite easily.

Remember that if you run aground once, it is likely to happen again as it indicates a particularly shallow stretch—or you are out of the channel.

In a town it is common to run aground on sunken rubbish, for example old oil drums, bicycle frames, bedsteads etc; this is most likely to occur near bridges and housing estates. Use the same methods, but be very careful as these hard objects can easily damage your boat or propeller.

Remember that winding holes are often silted up; do not go further in than you have to.

The difficulties mentioned above do not, of course, arise on the river Severn or the Gloucester & Sharpness Canal; however the BCN area requires extra care.

Mooring
All boats carry metal stakes and a mallet.

How a lock works

Plan: lock filling

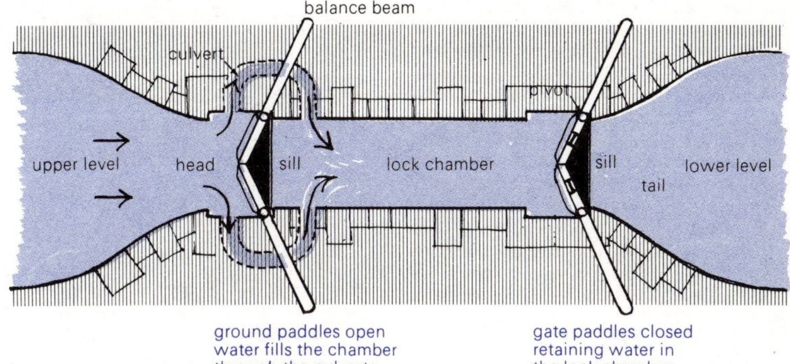

ground paddles open water fills the chamber through the culverts

gate paddles closed retaining water in the lock chamber

Elevation: lock emptying

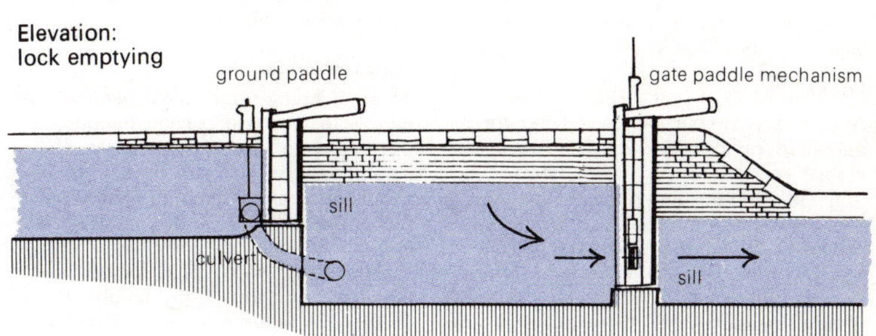

ground paddles closed preventing water from the upper level filling the chamber

gate paddles open water flows from the chamber to the lower level

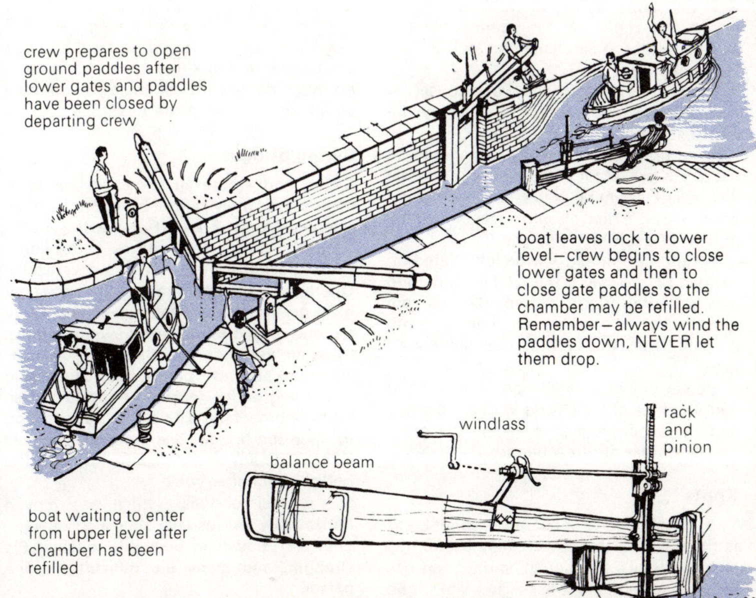

crew prepares to open ground paddles after lower gates and paddles have been closed by departing crew

boat leaves lock to lower level—crew begins to close lower gates and then to close gate paddles so the chamber may be refilled. Remember—always wind the paddles down, NEVER let them drop.

boat waiting to enter from upper level after chamber has been refilled

lock gate with paddle mechanism

These are used for mooring when there are no rings or bollards in sight, which is usually the case. Generally speaking you may moor anywhere to BWB property but there are certain basic rules. Avoid mooring anywhere that could cause an obstruction to other boats; do not moor on a bend or a narrow stretch, do not moor abreast boats already moored. Never moor in a lock, and do not be tempted to tie up in a tunnel or under a bridge if it is raining. Pick a stretch where there is a reasonable depth of water at the bank, otherwise the boat may bump and scrape the canal bed—an unpleasant sensation if you are trying to sleep. For reasons of peace and privacy it is obviously best to avoid main roads and railway lines.

Never stretch your mooring lines across the towpath; you may trip someone up and face a claim for damages. This means that you should never tie up to trees or fences.

There is no need to show a riding light at night, except on major rivers and busy commercial canals.

So long as you are sensible and keep to the rules, mooring can be a pleasant gesture of individuality.

Locks

A lock is a simple device, relying for its operation on gravity, water pressure and manpower.

On the preceding page, the plan (top) shows how the gates point uphill, the water pressure forcing them together. Water is flooding into the lock through the underground culverts that are operated by the ground paddles: when the lock is full, the 'top' gates (on the left in the drawing) can be opened. One may imagine a boat entering, the crew closing the gates and paddles after it.

In the elevation, the bottom paddles have been raised—opened—so the lock empties. A boat would of course float down with the water. When the lock is 'empty', the bottom gates can be opened and the descending boat can leave the lock.

Remember that:

1. For reasons of safety and water conservation, all gates and paddles must always be left closed when you leave a lock.
2. When going *up* a lock, a boat should be tied up to prevent it being thrown about by the rush of incoming water; but when going *down* a lock, a boat should never be tied up, or it will be left high and dry.
3. Windlasses should *not* be left slotted on the paddle spindle. If the ratchet slips (and they are often worn) the spindle will spin round and the windlass will fly off, probably into the lock or into someone's face.
4. Be very careful when operating locks in wet weather: the lockside is often slippery and the wooden planks across the gates can be downright treacherous.

Knots

You do not need to know much about knots as there is one that is generally useful, the clove hitch. This is simple, strong, and can be slipped on and off easily. Make two loops in a rope, and pass the right hand one over the left; then drop the whole thing over a bollard, post or stake and pull it tight. See diagram below.

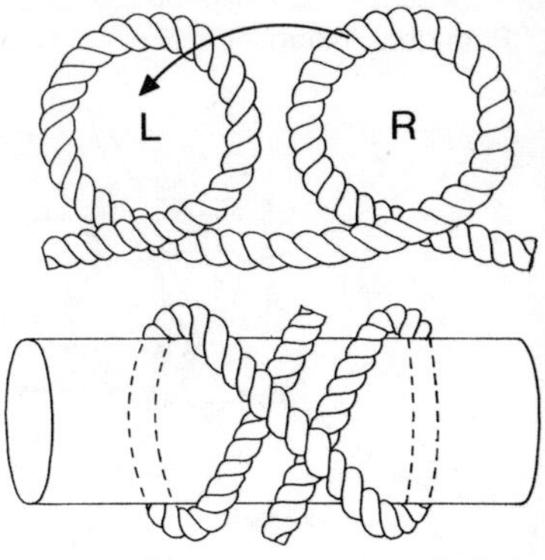

When leaving a mooring coil all the ropes up again. They will then be out of the way, but ready if needed in a hurry. Many a sailor has fallen overboard after tripping on an uncoiled rope.

Fixed bridges

At most bridges the canal becomes very narrow, a means of saving building costs developed by the engineers. As a result careful navigation is called for if you are to avoid hitting either the bridge sides with the hull or the arch with the cabin top. As when entering a lock, the best way to tackle 'bridgeholes' is to slow down well in advance and aim to go straight through, keeping a steady course. Adjustments should be kept to a minimum for it is easy to start the boat zig-zagging, which will inevitably end in a collision. One technique is to gauge the width of the approaching bridgehole relative to the width of the boat, and then watch one side only, aiming to miss that side by a small margin—say 6in; the smaller you can make the margin, the less chance you have of hitting the other side of the bridge. If you do hit the bridge sides when going slowly it is not likely to do much damage; it will merely strengthen your resolve to do better next time.

Swing bridges

Swing bridges are an attractive feature of some canals; they cannot be ignored as they often rest only 2 or 3ft above the water. Generally they are moved by being either swivelled horizontally, or raised vertically. Operation is usually manual, although some have gearing to ease the movement. There are one or two mechanized swing bridges; these are very rare, and they have clear instructions at control points. Before operating any swing bridge make sure that any road traffic approaching is aware of your intention to open the bridge. Use protective barriers if there are any and remember to close the bridge again after you.

Swivel bridges, which are moved horizontally, are usually simple to operate. They do however demand considerable strength, and many are difficult for one person.

Lift bridges, which are moved vertically, are raised by pulling down a balance beam. The heaviest member of the crew should swing on the chain that hangs from the beam. Once the bridge is up the beam should be sat on or otherwise held, as in some cases it fails to counterbalance the bridge. Serious damage could be caused to the boat and to the helmsman if the bridge were allowed to fall while the boat was passing through. A draw bridge is another, more traditional type of lift bridge; the method of operation is the same.

Tunnels

Many people consider a canal incomplete without one or two tunnels, and certainly they are an exciting feature of any trip. Nearly all are easy to navigate, although there are a few basic rules:

Make sure your boat has a good headlight in working order.

If you can see another boat in the tunnel coming towards you, it is best to wait until it is out before entering yourself. It is in fact possible for craft of 7ft beam to pass in many tunnels, but it can be unnerving to meet another boat. If you do, keep to the right as usual.

In most tunnels the roof drips constantly, especially under ventilation shafts. Put on a raincoat and some form of hat before going in.

A notice on the tunnel portal will give its length, in yards, and will say whether unpowered craft are permitted to use it.

Care of the engine

Canal boats are powered by one of three types of engine; diesel, petrol and petrol/oil or two-stroke. However basic rules apply to all three. Every day before starting off, you should:

Check the oil level in the engine.
Check the fuel level in the tank.

If your engine is water-cooled, check that the filter near the intake is clean and weedfree. Otherwise the engine will overheat which could cause serious damage.

Check the level of distilled water in the battery, and ensure that it is charging correctly.

Lubricate any parts of the engine, gearbox or steering that need daily attention.

Check that the propeller is free of weeds, wire, plastic bags and any other rubbish. Although this is an unpleasant task, it is a constant necessity and will remain so as long as canals continue to be used as public rubbish dumps. The propeller and the water filter should be checked whenever there is any suspicion of obstruction or overheating—which may mean several times a day.

When navigating in shallow water, keep in mind the exposed position of the propeller. If you hit any underwater obstruction put the engine into neutral immediately. When running over any large floating object put the engine into neutral and wait for the object to appear astern before re-engaging the drive.

Respect should be shown to the engine. Remember that this simple maintenance could make the difference between trouble-free cruising and tiresome breakdowns.

Fuel

Petrol engines and petrol/oil motors are catered for by some boatyards and all roadside fuel stations. Fuel stations on roads near the canal are shown in the guide, and these should be considered when planning your day's cruise. Running out is inconvenient; remember you may have to walk several miles carrying a heavy can.

Diesel powered boats pose more of a problem in obtaining fuel, although their range is generally greater than that of petrol powered craft. Most boatyards sell marine diesel, which is tax-free and therefore very much cheaper (the heavy tax on fuels is aimed only at road vehicles). The tax-free diesel, which contains a tell-tale pink dye to prevent it being used in road vehicles, can only be sold by boatyards, which are still few and far between on the canals. BWB yards may let you have enough to reach the next boatyard. Some, but by no means all roadside fuel stations sell diesel, but at the higher price. A further complication is that diesel engines must not be allowed to run out of fuel, as their fuel system will need professional attention before they can run again. The simple rule about fuel is—think ahead. It is advisable to carry a spare can.

Water

Fresh water taps occur irregularly along the canals, usually at boatyards, BWB depots, or by lock cottages. These are marked on the maps in the guide. Ensure that there is a long water hose on the boat (BWB taps have a $\frac{1}{2}$-inch slip-on hose connection).

Lavatories

Canal boats are generally fitted with chemical lavatories which have to be emptied from time to time. Never empty them over the side or just tip them into the bushes. Either empty them at the sewage disposal points marked on the maps, or dig deep holes with a spade. Some BWB depots and boat yards have lavatories for boat crews.

Litter

Some canals are in a poor state today because they have long been misused as unofficial dumps for rubbish, especially in towns. Out of sight is out of mind only until some object is tangled round your propeller. So keep all rubbish until you reach a refuse disposal point. (See the maps).

BWB Byelaws

Although no one needs a 'driving licence to navigate a boat on the waterways, boat users should remember that they incur certain responsibilities and duties to the BWB and to other waterway users, e.g. a knowledge of correct sound signals. These liabilities are contained in the BWB Byelaws, which should be obtained by all prospective navigators by sending 12½p to the Secretary, British Waterways Board, Melbury House, Melbury Terrace, London NW1 6JX.

Practical information fishing

This section has been written by Bill Howes, well known angling correspondent and author of many books on the subject.

Most of these cross-country waterways have natural reed-fringed and grassy banks, and in addition to the delightful surroundings the fishing is generally good. In most areas there has been a steady improvement in the canal fishing in recent years and in many places new stocks of fish have been introduced. Good stocks of quality bream and roach have gone into several canals in the past couple of years, and this has improved the sport considerably.

The popular quarry are roach, perch and bream, but the canals also hold dace, tench, chub and carp in places, in addition to pike and other species in particular areas.

Canals afford good hunting grounds for those seeking specimen fish, that is fish above average size, and these are liable to be encountered on almost any water, yet some areas are more noted for big fish than others. The canals also make good venues for competition fishing, and in most places nowadays matches are held regularly at week-ends throughout the season.

The Statutory Close Season for coarse fish is March 15 to June 15 inclusive, but in some areas, the Close Season is from February 28 to May 31. The Close Season for pike in some areas is March 15 to September 30.

Permits and fishing rights

Most parts of the canal system are available to anglers. Several angling associations in the South West—Birmingham AA, Gloucester AA, Gwent & Border Counties AS, Newport AA, Pontypool AA and Worcester & District United AA—rent fishing rights from the BWB over extensive areas on the canal system. In most cases day-tickets are available.

On arrival at the canal-side it is always advisable to make enquiries as to who holds the fishing rights, and to obtain a permit if one is required *before* starting to fish.

Remember, also, that a River Authority rod licence is required in addition to a fishing permit. It is essential to obtain this licence from the relevant River Authority *before* starting to fish. Some fishing permits and licences are issued by bailiffs along the bank, but local enquiry will help to determine this.

A canalside pub or a local fishing tackle shop are good places to enquire if permission or day tickets are required for the local stretch of water. Canal lock-keepers are usually knowledgeable about the fishing rights in the immediate locality, and often a lock-keeper may be found who issues day tickets on behalf of an angling association, or owner.

A lock-keeper is always worth talking to, for apart from knowing who holds the fishing rights, he often knows some of the better fishing areas, as well as local methods and baits which may be considered most successful.

The fishing rights on most canals are owned by the British Waterways Board and many miles of good fishing are leased to clubs and angling associations. They also issue day tickets on certain lengths, so it is worth enquiring at the local British Waterways office when planning a trip. Special arrangements are made for fishing from boats. Consult BWB at Nantwich (65122).

Once permission has been obtained it would be advisable to find out if there are any restrictions imposed, since some clubs and associations ban certain baits, or have restrictions on live-baiting for pike; and on some fisheries pike fishing is not allowed before a specified date.

Other restrictions may concern size-limits of fish: some River Authority bye-laws prohibit the retention of under-sized fish in keep nets. A local club holding the fishing rights might have imposed their own size-limits in order to protect certain species. Such restrictions are generally printed on permits and licences.

Tackle

In the slow moving, sluggish waters the float tackle needs to be light and lines fine in order to catch fish.

When fishing for roach and dace lines of $1\frac{1}{2}$lb to 2lb breaking strain are the maximum strength normally needed in order to get the fish to take a bait—particularly when the water is clear, or on the popular reaches which are 'hard-fished'.

Fine tackle also means small hooks, and hook sizes 16 and 18—or even as small as 22 at times. Such light gear is also effective when fishing for the smaller species, such as gudgeon and bleak. This tackle will require a well-balanced float to show the slightest indication of a bite.

Bait

Baits should be small, and maggots, casters (maggot chrysalis), hempseed, wheat, tiny cubes of bread crust, or a small pinch of flake (the white crumb of a new loaf) may take fish.

But it pays to experiment with baits; bait which is effective on one occasion will not necessarily prove to be as effective the next. With slight variations, similar fishing methods can be used effectively on the majority of waterways.

Northern anglers who regularly compete in contests on canals use bloodworms as bait. They have become extremely skilful in using this tiny bait and often take fish on bloodworms when all other baits fail. Bloodworms are the larvae of a midge, and are a perfectly natural bait. The anglers gather the bloodworms from the mud and, apart from a wash in clean water, the baits are ready for use.

A popular groundbait which has had great success is known as 'black magic'! This is a mixture of garden peat and bread crumbs mixed dry and carried to the water. When dampened and mixed it can be thrown in in the usual way.

The basis of most groundbaits is bread, and many other materials may be added. But always avoid stodgy mixtures for canal fishing. Canals are not waters which respond to heavy groundbaiting tactics.

It is far better to use a cloud-bait, and this can be purchased ready for use. Some successful midland anglers wet their cloud-bait with milk instead of water to increase the cloud effect.

Methods

Once the swim–that is the area of water to be fished–has been decided upon, and the tackle set up, use the plummet to find the depth and adjust the float.

It usually pays to plumb the depth of the swim before fishing, but be cautious when doing so in clear waters. At times it may be best to find the depth by trial and error.

Often most fish will be caught from around mid-water level, but always be prepared to move the float further up the line in order to present the bait closer to the bottom, where the bigger fish are usually to be found.

At frequent intervals toss a few samples of the hook-bait into the top of the swim to keep the fish interested.

Fish vary in different swims, and on different waters, in the way they take a bait and this creates a different bite registration. It may be found that the fish take the hook-bait quickly, causing the float to dip sharply or dive under the surface. The strike should be made instantly, on the downward movement. On some canals the fish are even quicker–and perhaps gentler–not taking the float under at all, and in this case the strike should be made at the slightest unusual movement of the float.

Roach and dace abound in many lengths and although working the float tackle down with a flow of water takes most fish, better quality fish–also bream–are usually to be taken by fishing a laying-on style, with the bait lying on the bottom. This method can often be best when fishing areas where there is no flow at all.

This can be done with float tackle, adjusted to make the distance from float to hook greater than the depth of water, so that when the float is at the surface the bait and lower length of line are lying on the bottom.

The alternative method of fishing the bottom is by legering, the main difference in the methods being in the bite indication. Without the float a bite is registered at the rod-tip where, if need be, a quiver-tip or swing-tip may be fitted. These bite detectors are used extensively on Midland and Northern waters.

Legering is a method often used in the south, where in some southern canals barbel and chub are quite prolific. These species grow to good sizes in canal waters–chub up to 7lb and barbel up to 14lb have been taken–but these are exceptional and the average run of fish would be well below those weights. Nevertheless, both species are big fish and big baits and hooks may be used when fishing for them.

Many bigger than average fish–of all species–have been taken by fishing the bait on the bottom. Whatever the style of leger fishing, always choose the lightest possible lead weight, and position it some 12 to 18 inches up from the hook. There are no hard and fast rules governing the distance between lead and hook, so it pays to experiment to find the best to suit the conditions.

Anglers who regularly fish the northern, and midland canals invariably use tiny size 20 and 22 hooks, tied to a mere $\frac{3}{4}$lb b.s. line, and when float-fishing use a tiny quill float–porcupine or crow quill. A piece of peacock quill is useful because it can be cut with scissors to make it suit prevailing conditions.

Such small floats only need a couple of dust-shot to balance them correctly, and usually the midland anglers position this shot on the line just under the float so that the bait is presented naturally. Once the tackle has been cast out the bait falls slowly through the water along with hook-bait samples, which are thrown in at the same time. This is called 'fishing on the drop'. A fine cloud-bait is also used with this style.

Canals which have luxuriant weed growth harbour many small fish, which are preyed upon by perch. These move in shoals and invariably the perch in a shoal are much the same size. Usually the really big perch are solitary, so it pays to rove the canal and search for them.

Perch are to be caught from almost any canal and although they may be caught by most angling methods, the most effective is usually float-fishing. The fishing depth can vary according to conditions, time of year, and actual depth of the canal, so it pays to try the bait at varying depths. The usual baits for perch are worms, small live-baits (minnows etc) and maggots. Close by the wooden lock gates are very often good haunts for perch.

In certain places canals and rivers come together and take on the characteristics of the river (i.e. with an increased flow) and different methods are needed. These places are often noted for splendid chub (and sometimes barbel) in addition to roach and other species. Trotting the stream is a popular and effective fishing style.

Weather

Weather conditions also have to be taken into consideration. Canals usually run through open country and catch the slightest breeze. Even a moderate wind will pull and bob the float, which in turn will agitate the baited hook. If bites are not forthcoming under such conditions then it may be best to remove the float and try a straight-forward leger arrangement.

When legering, the effects of the wind can be avoided by keeping the rod top down to within an inch or two of the water level—or even by sinking the rod-tip below the surface. Anglers in the north and mid-lands have devised a wind-shield for legering which protects the rod-tip from the wind and improves bite detection.

Nevertheless, in some circumstances a slight wind can be helpful because if a moderate breeze is blowing it will put a ripple on the water, and this can be of assistance in fishing in clear waters.

Where to fish

Most canals are narrow and this makes it possible to cast the tackle towards the far bank where fish have moved because they had been disturbed from the near bank. Disturbance will send the fish up or down-stream and often well away from the fishing area. So always approach the water quietly, and remember to move cautiously at all times. When making up the tackle to start fishing it is advisable to do so as far back from the water as possible to avoid scaring the fish. It pays to move slowly, to keep as far from the bank as possible, and to avoid clumping around in heavy rubber boots.

If there is cover along the bank—shrubs, bushes, tall reeds and clumps of yellow flag iris—the wise angler will make full use of it.

There are some canals which are not navigable, and these are generally weedy. At certain times in the season the surface of the water disappears under a green mantle of floating duckweed, which affords cover and security for the fish. It is possible to have the best sport by fishing in the pockets of clear water which are to be found.

Some canals have prolific growths of water lilies in places, and are particularly attractive for angling. They always look ideal haunts for tench, but they can also be rather difficult places from which to land good fish.

Tench are more or less evenly distributed throughout the canals and the best are found where weed growth is profuse. It may be best to fish small areas of clear water between the weeds. Groundbait can encourage tench to move out from the weed beds, and to feed once they are out. Sometimes it is an advantage to clear a swim by dragging out weeds or raking the bottom. This form of natural groundbaiting stirs the silt which clouds the water, and disturbs aquatic creatures on which the fish feed. Sometimes a tench is hooked within minutes of raking.

Bream seem to do well in canals and some fairly good fish up to 5lb may be taken. Some canals are noted for shoals of big bream, where it is possible to catch over 50lb of them at a session. Any deep pools or winding holes (shown as ∩ on map) are good places to try, particularly when fishing a canal for the first time.

Other places worth fishing are 'cattle drinks' regularly used by farm animals. These make useful places to fish for bream, roach and dace. The frequent use of these drinking holes colours the water, as the animals stir up the mud, and disturb various water creatures. The coloured water draws fish into the area—on the downstream side of the cattle drink when there is the slightest flow.

Pike are to be found in every canal in the country, and these grow big. They are predators, feeding on small fish (which gives a sure indication of the most effective baits.) Any small live fish presented on float tackle will take pike. The best places to fish are near weed beds and boats which have been moored in one place a long time.

As a general rule, never fish in locks on navigable canals, or anywhere that could obstruct the free passage of boats. Remember that you will inconvenience yourself as well as the boatman if you have to move in a hurry, or risk a broken line. And please respect the enjoyment of others: do not obstruct the towpath with gear or leave litter of any kind.

British Waterways Board
The BWB Fisheries Officer at Watford welcomes specific enquiries about fishing on BWB canals from individuals, associations and clubs. He will also supply the name and address of the current Secretary of each Angling Association shown on page 18.

River Authorities
Refer to the map on pages 4 and 5 showing the areas covered

Avon & Dorset River Authority
3 St. Stephens road,
Bournemouth, Hants. (24700).
Rod licence required.

Bristol Avon River Authority
Green Park road,
Bath, Somerset. (27541).
Rod licence required.

Severn River Authority
Portland House,
Church street,
Great Malvern,
Worcs. (61511).
Rod licence required.

Somerset River Authority
The Watergate,
West Quay, Bridgwater,
Somerset. (8271).
Rod licence required.

Trent River Authority
206 Derby road,
Nottingham. (42300).
Rod licence required.

Usk River Authority
The Croft,
Coldcross Common,
Caerleon, Mon. (420399).
Rod licence NOT required.

General information

Canal walks

Since virtually every artificial canal in England was originally designed to carry horse-drawn boats, a continuous towing path was usually constructed along the entire length of the navigation. Nowadays this path is hardly ever used for towing, and thus forms a splendid long-distance walking route which combines all the fascination of canals with their valuable seclusion from all motor vehicles. However the gradual disappearance of horse-drawn boats and the constant erosion of canal banks by the wash of motor boats has meant that in places the towpath has become narrow and difficult to negotiate. At the same time the towpath hedge often manages to outstrip efforts to keep it trimmed and tidy, and for these reasons a pair of waterproof boots is a useful item on a canal walk. Remember also that canal towpaths are not necessarily public rights of way. (Maps showing rights of way can be inspected at Local Authority offices.) The British Waterways Board are willing to arrange with Local Authorities for the public to be given access to all the towing paths and for their condition to be improved.

River walks

River navigations are not usually well suited to long distance walking. There is often no towpath as such—the old navigation companies generally had to obtain towing rights from the land owners whose property fronted on to the river. Where the towing rights switched to the other side of the river without a bridge, the towing horse would get into its boat, which would then be pushed across to the opposite side. This solution is unfortunately not normally available to walkers, who have to retrace their steps to the last bridge. The presence of weirs—usually unbridged—adds to the difficulties of walking far along river navigations. While the maps in this guide are intended to give as realistic a representation as possible, the depiction of a towpath beside a navigation does not guarantee the presence on the ground of a good footpath.

Planning walks

Although rural bus services are constantly being pruned, many survive and provide useful access to the waterways. Dozens of the places described in the guide are linked by bus services, so it is usually possible to plan a one-way walk.

Similarly, railway stations continue to serve the walker in many parts of the country. (A set of BR timetables is a very useful thing to keep on canal boats to facilitate train/boat link-ups). Local enquiry on bus and train services is often rewarding. In the last resort, the village taxi is available.

Motoring

Since most canal bridges are still the old-fashioned hump-backed variety, persons visiting canals by motor car should remember to park their machine well away from the bridge to avoid obstructing other traffic.

Finally, all readers are reminded that at the moment the canals are unspoilt, peaceful and greatly appreciated by all who use them. It is to be hoped that they remain so.

A country code for waterway users

Outdoor recreation of all sorts is increasingly popular and is creating unaccustomed pressures in our countryside. The greater use of waterways is just one such pressure, but there is no need for conflict between those who seek enjoyment in the countryside and those who make their living from it. A little common sense and courtesy go a long way.

Every waterway user will want to make use of the banks or towpaths at some stage, but if you want to picnic or camp on land other than the towpath beside a waterway, remember to ask the farmer first.

Walking in the countryside is one of the cheapest and most enjoyable forms of recreation, but all farm crops will suffer from indiscriminate trampling—so always keep to paths across farm land.

Dogs are a special problem and can all too often frighten or worry livestock: sheep are particularly vulnerable. If you have a dog, make sure it is under control.

Field boundaries, whether they are hedges, fences or walls, have to be kept in good condition to prevent livestock from straying. Be careful not to damage them.

Litter in the countryside is always unpleasant and often dangerous. Boatmen should keep it on board until reaching a BWB disposal point. If you are an angler, take your litter home and do not leave broken lines and hooks lying around; they may injure an animal or bird.

During the summer, remember the risk of fire, particularly with woodland, ripe corn or straw.

Protect wildlife, plants and trees.

Marsh Tit
Nuthatch
Snipe
Oystercatcher
Blackbird
Great Crested Grebe
Bullfinch
Mistle Thrush
Shoveler
Kestrel
Quail
Tawny Owl
Magpie
Treecreeper
Wren
Great Spotted Woodpecker
Spotted Flycatcher
Curlew
Great Tit

Kennet & Avon

Maximum dimensions

Reading to Bath, junction with river Avon
Length: 72′ or 70′
Beam: at 7′ at 13′ 9″
Headroom: 7′ 6″
Bath to Hanham lock
Length: 75′
Beam: 16′
Headroom: 8′ 9″

Mileage

READING to
Aldermaston Wharf: 10
Tile Mill lock: 8
Aldermaston Wharf: 10
Newbury lock: 18½
Kintbury: 24½
Hungerford: 27½
Crofton top lock: 35
Pewsey Wharf: 41½
Devizes top lock: 53½
Bradford-on-Avon: 65½
Dundas Aqueduct: 70
Bath, junction with river Avon: 75¼
HANHAM lock (start of tidal section): 86½
Bristol Docks: 93
AVONMOUTH entrance to Severn Estuary: 100¾

Total 107 locks

The Kennet & Avon Canal is one of the most splendid lengths of artificial waterway in Britain, a fitting memorial to the canal age as a whole. It is a broad canal, cutting across southern England from Reading to Bristol. Its generous dimensions and handsome architecture blend well with the rolling downs and open plains that it passes through, and are a good reminder of the instinctive feeling for scale that characterised most 18thC and early 19thC civil engineering.

The canal was built in three sections. The first two were river navigations, the Kennet from Reading to Newbury, and the Avon from Bath to Bristol, both being canalised. Among early 18thC river navigations the Kennet was one of the most ambitious, owing to the steep fall of the river. Between Reading and Newbury 18 locks were necessary in as many miles, as the difference in level is 138 feet. John Hore was the engineer for the Kennet Navigation, which was built between 1718 and 1723 and included 11 miles of new cut. Subsequently Hore was in charge of the Bristol Avon Navigation, carried out between 1725 and 1727. These river navigations were interesting in many ways, often because of the varied nature of the country they passed through. The steep-sided Avon gorge meant that a fast-flowing river had to be brought under control. Elsewhere the engineering was unusual: for example the turf sided locks on the Kennet, some of which still survive.

For the third, linking, stage a canal from Newbury to Bath was authorised in 1794. Rennie was appointed engineer, and after a long struggle the canal was opened in 1810, completing a through route from London to Bristol. The canal is 57 miles long, and includes 79 broad locks, a summit level at Savernake 474 feet above sea level and one short tunnel, also at Savernake. Rennie was both engineer and architect, anticipating the role played by Brunel in the creation of the Great Western Railway; in some ways his architecture is the more noteworthy aspect of his work. The architectural quality of the whole canal is exceptional, from the straightforward stone bridges to the magnificent neo-classical aqueducts at Avoncliff and Limpley

Stoke. Rennie's engineering, however, left something to be desired; the summit level was too short, and so pumping stations had to be installed at Crofton and Claverton to maintain the water level and feed the locks; in other places the canal bed was built over porous rock, and so leaked constantly, necessitating further regular pumping.

Nevertheless the canal as a whole was a striking achievement. At Devizes the canal descends Caen Hill in a straight flight of 29 locks, the longest flight of broad locks anywhere in Britain. The many swing bridges were designed to run on ball bearings, one of the first applications of the principle. The bold entry of the canal into Bath, a sweeping descent round the south of the city, is a firm expression of the belief that major engineering works should contribute to the landscape, whether urban or rural, instead of imposing themselves upon it, as often happens nowadays.

Later the Kennet and Avon Canal Company took over the two river navigations, thus gaining control of the whole through route. However traffic was never as heavy as the promoters had expected, and so the canal declined steadily throughout the 19thC. It suffered from early railway competition as the Great Western Railway duplicated its route, and was eventually bought by that railway company. Maintenance standards slipped, and this, combined with a rapidly declining traffic, meant that navigation was difficult in places by the end of the 1914–18 war. The last regular traffic left the canal in the 1930s, but still it remained open, and the last through passage was made in about 1947. Subsequently the canal was closed, and for a long time its future was in jeopardy. However, great interest in the canal had resulted in the formation of a Canal Association shortly after the 1939–45 war to fight for restoration. In 1962 the Kennet & Avon Canal Trust was formed out of the Association, and practical steps towards restoration were under way. Using volunteer labour, funds raised from all sources, and with steadily increasing help from BWB, the Trust have reopened the canal westwards from Newbury towards Hungerford, have restored the Wootton Rivers flight of locks to increase the navigation of the long pound, and are reopening the Bath flight. Although the major work is still ahead, the Trust have proved that the reopening of the Kennet & Avon is both practical and desirable. Within a few years boats may once again be able to cruise across Britain from London to Bristol. There is a public right of way along the towpath, making this canal a magnificent long distance walk.

Natural history

This canal forms a unique series of freshwater habitats which vary along its length according to depth of water, aspect, height of the banks and the time of the year. The water is deepest to the east of the main source of water—Wilton Water (near Crofton), itself an excellent wildlife habitat—and remains fairly constant to the west, within seasonal variations, as far as the top lock at Devizes. From the bottom of the flight of locks the water is shallower, with much silting in some places, until it reaches Bradford-on-Avon. It is dry from here to Limpley Stoke. The last section, from Bathampton to Bath, where it joins the River Avon, is one of the most interesting to naturalists, because the regular passage of the 'Jane Austen' down the centre leaves untouched a belt of shallow water on both sides. The marginal and semi-marginal plants, including the flowering rush, are exceptionally fine.

The stonework of bridges, locks and aqueducts provides additional habitats for shallow-rooting plants. The three ferns, hartstongue, wall-rue and black spleenwort are of special interest. Where the lock walls are capped with limestone, lime-loving plants such as fairy flax and quaking-grass are found.

For much of its length the canal is bordered on one or both banks by trees, including several species of willow, and bushes, predominantly hawthorn. A detailed survey in 1972 showed that plants along the towpath vary little, growing in profusion along its length. There are many kinds of grasses; other frequent species are white deadnettle, hogweed, meadow cranesbill, ground ivy and fine specimens of the ratstail plantain. Apart from the marsh marigold, the yellow iris and the greater pond sedge, there are few flowers at the water's edge and on the canal bank until after mid-summer when purple loosestrife, the three-petalled arrowhead and flowering rush, together with the clusters of small creamy-white flowers of meadowsweet and pink-tinged angelica, contrast with the tall reed-grass and reedmace (often mistakenly called 'bulrush'). Rooting just in the water, the commonest plant is the branched bur-reed. It has a profusion of leaves but the green, spherical, prickly-looking flowers are often sparse. Large tufts of tussock sedge, clumps of water dock and short stretches of common reed, having all the leaves turning on the tall stems to face away from the wind, can be found at the water's edge. Of all the marginal plants, perhaps the most handsome is the flowering rush which is commonest near Bath. Deeper water supports such plants as yellow water-lily, water crowfoot and several pondweeds which root in the mud but hold their leaves and flowers above the water surface. A few plants are entirely submerged, the commonest being hornwort which grows in long bushy, brittle tassels.

In the summer and autumn much of the water surface is covered with small floating plants which are moved by the wind. There are four species of duckweed and a water fern, azolla, accidentally introduced from North America, which turns the canal red in autumn. It has become a problem near Seend where, with the duckweeds and blanket weed, it forms a blanket 4 inches thick, eliminating light and air from the water surface. In hot weather this causes hundreds of fish to suffocate. The weed is only killed by a hard frost, when it sinks to the bottom and decays.

Between mid-April and mid-August 1972 a botanist recorded 190 different species of plants associated with the presence of the canal along a 4 mile stretch near Melksham. The different plant zones make not only for numerical and visual diversity but also for the diversity of animals dependent on them. Mute swans, coots and moorhens nest and take refuge in the dense vegetation at the water's edge, water voles feed on it and use rushes and grass to line their nests. These are four of the canal's largest animals but they, like all wild life, are part of the food chains which depend on the oxygen in the water and the light that filters into it. Amongst the tangle of submerged green plants and their roots lives a host of small animals, many visible only under a microscope. Some of the water snails, the large swan mussels (8 inches long), fresh-water shrimps and flatworms feed on decaying plant matter, while others such as water beetles, water spiders, water boatmen, dragonfly nymphs and the strange water stick insects, represent the hunting predators, actively seeking out their quarry which is often much larger than themselves. Many of these freshwater creatures are found only in the comparatively still water of canals and ponds for which they are adapted for living, for they would be swept away in the streams and rivers. They provide food for water shrews, fish, frogs and, occasionally, grass snakes which can be seen on sunny days curled up on the towpath or swimming across the canal in a few regular haunts.

The shy little grebe, with its trilling call, has spread from Wilton Water and now nests, on platforms made by willow branches dipping into the water, as far west as Devizes. Kingfishers have returned to the localities they frequented prior to the severe winters of the early 1960s and single herons are regular fishermen along the canal, though their diet also includes frogs, etc. The abundance of newly-emerged insects attracts many small birds to the canal. In spring, the bushes are sometimes full of chiffchaffs and willow warblers eagerly feeding after their long flight from their winter quarters. Later, sedge and reed warblers arrive, take up their territories along the canal, sometimes singing all day and night, and build their nests in rushes and reeds. Swallows and house martins leave their nesting sites in the villages on summer evenings to feed over the canal, skimming low to pick up insects off the surface of the water. In late summer brambles growing on the banks are then in flower attracting butterflies to feed on them. The commonest are meadow brown, speckled wood, small tortoiseshell, peacock, brimstone and the quaintly-named gatekeeper. In autumn, pied wagtails converge on the canal pounds near Devizes at dusk to roost in the rushes. By December they number around 300. At one point on the canal, the bank vegetation dies down; in winter a huge badger sett, excavated in the high greensand canal bank, can be seen. Here, the only length of the towpath that can comfortably be walked is the 50 yards used by the badgers on their nightly journeys to and from their feeding grounds.

The canal offers a wealth of interest and for students of all ages it can serve as an open-air natural history laboratory. If restoration work succeeds in producing the healthy ecological balance that has emerged near Bath, then naturalists, anglers and all other users will benefit.

A contemporary view of canal promoters. *Eric de Maré.*

Above: Caen Hill locks, Devizes, summer 1972.
Below: Kintbury lock after restoration, summer 1972.

Reading

4 miles

The river Kennet leaves the Thames east of Reading. The mouth of the river is marked by two large gasholders, a splendid gleaming, hissing, modern gasworks and the main railway which runs parallel to the south bank of the Thames. The Kennet leads south west towards the centre of Reading, passing Blakes lock, the only lock maintained by the Thames Conservancy that is not actually on the Thames. The Kennet through Reading is narrow, shallow and fast-flowing, being a river navigation; also there are several sharp blind bends in the town so great care in navigation is needed. Keep a sharp lookout for other boats and remember to allow for the flow of the river. The river cuts across the middle of the town, and so access to all facilities is easy. Rows of riverside cottages and a surprising variety of bridges decorate the Kennet in Reading, London street bridge being the most central access point. The Kennet passes over a weir with County Lock adjoining. The weirs are a feature of river navigation that should be treated with respect, as the current they create can often affect the course of a boat, especially when making a slow approach to a lock. The river gradually leaves the town, passing through Fobney Meadow to Fobney lock. The more open countryside continues to Southcote lock.

Navigational note
The Kennet & Avon Canal locks require windlasses of an intermediate gauge – $1\frac{1}{8}$in. instead of the normal 1in. or $1\frac{1}{4}$in. Such windlasses may be bought or hired from the Reading Marine Company or from the BWB at Padworth (*see page 26*).

Reading
Berks. Pop 124,000. EC Mon/Wed. MD Mon. The town lies at the extremity of the Berkshire Downs and the Chiltern Hills, where the Thames becomes a major river. It is the Victorian architecture that makes this town interesting, as the university buildings are not to everyone's taste. Canal walkers in Reading will find there is no towing path in the centre of town; however west of Reading the whole canal is a public right of way.

Abbey Ruins Fragmentary remains of this 12thC abbey built by Henry I lie on the edge of Forbury Park. The 13thC gatehouse, altered by Scott in 1869, still stands.

The Gaol Forbury rd. Designed by Scott and Moffat in 1842–4 in the Scottish baronial style. Oscar Wilde wrote his 'Ballad of Reading Gaol' while imprisoned here.

Museum of English Rural Life Whiteknights Park. A fascinating collection of relics of old English agriculture, and a small display of painted canal ware with a set of tools used for making narrow-boats. *Open Tue-Sat 10.00–13.00, 14.00–16.30. Closed Sun, Mon, B. Hols.* (Reading 85123, ext. 475).

Museum & Art Gallery Friar st. Has an exceptional natural history and local archaeology collection. Also prehistoric collection. *Open weekdays 10.00–17.30.*

BOATYARDS & BWB

Ⓑ **Kennet Mouth Boatyard** Kennetside, Reading. (64186). Moorings, repairs, water, petrol, slipway, chandlery, hire cruisers.

Ⓑ **Reading Marine Co.** Crane wharf, Kings rd, Reading. (53917). Moorings, repairs, water, petrol (50yds), slipway, chandlery, gas. K & A-sized windlasses for sale or hire.

PUBS

Cooper's Wine Lodge 29 Market place, Reading. (52238). Food (lunch only). Occasional pub entertainment.

Capri Restaurant 62 Christchurch rd, Reading. (82823). Restaurant (Italian), reductions for children and parties. *Closed Sun.*

Fisherman's Cottage Kennetside, Reading. Canalside. Food.

Jolly Anglers Kennetside, Reading. (56229). Canalside. Food, water.
Thames Tavern Kennetside, Reading. Canalside. Food.

FISHING

The *Kennet mouth at Caversham* is noted for roach, dace, chub, in fact all Thames species of fish. A huge 18-lb barbel was once caught at the mouth of the Kennet—one of the largest ever caught in the British Isles. That was many years ago, yet (smaller!) barbel are still to be found here.
Near *County lock, Reading* has always been exceptionally good for dace. Other species include roach, and barbel to 12lb in recent years.
From *Reading to Fobney* the fishing is first class. Good roach and dace, plus chub and superb barbel. There are some big bream and once a shoal has been located a big catch can be made. Carp to 10lb have also been caught. A barbel of 14lb from Fobney was reported in the angling press a few years ago.
At *Sulhamstead* there are some good tench, plus roach, dace and barbel. The best chub taken here weighed over 6lb. Good fish of all species are to be taken throughout this waterway.
The BWB has good fishing facilities available to anglers on a day ticket basis at its fishery on the Kennet & Avon canal at Sulhamstead, permits to fish the water being obtainable from the bailiff (Mrs. A. Bartlett) at Canal Cottage, Sulhamstead, Reading (302370). The fishery is capable of accommodating small parties of anglers or angling clubs (up to 30 members) and the venue can be reserved for match-fishing purposes etc. by contacting Mrs. Bartlett direct.
Burghfield has many fine swims which hold most species. Barbel of around 10lb, chub around 5lb and good tench have been caught. There are some fair-sized pike to be caught in winter.
Theale also yields excellent fishing for most Kennet species. Particularly good are the roach, dace, chub and barbel, but other kinds of fish swell the total bags of fish taken by club anglers.

TOURIST INFORMATION CENTRE

Reading Central Library, Blagrave st, Reading. (55911 ext. 484, or 54382).

Pewsey Wharf, on the 15-mile pound between Wootton Rivers and Devizes.

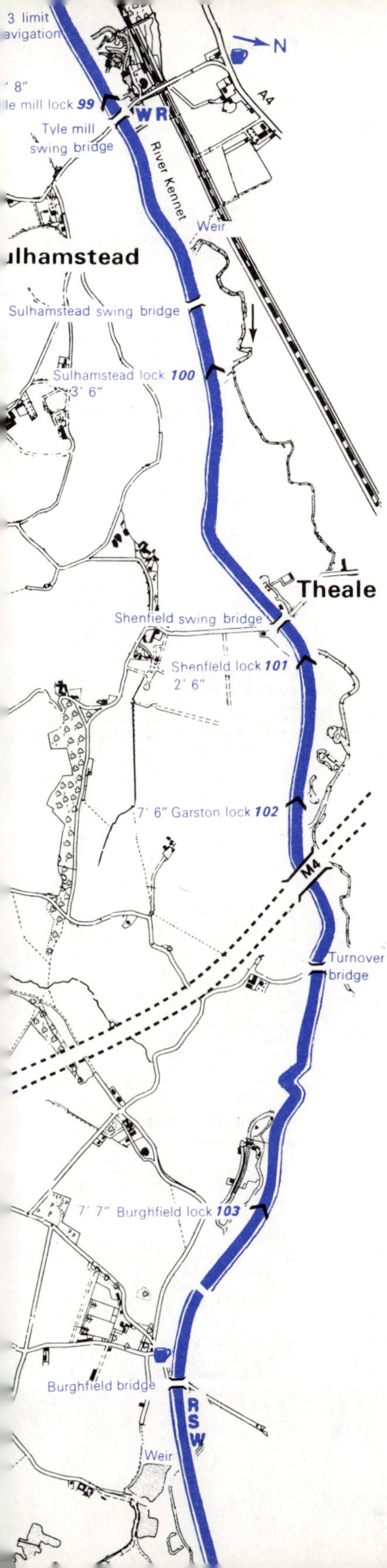

Theale

5 miles

Continuing west, the canal passes Burghfield bridge, a handsome stone arch. Burghfield lock is turf-sided, one of several that survive from the early Kennet navigation. This old style of lock makes getting on and off the boat slightly more difficult. The Kennet winds through water meadows, the straight stretches marking the canal sections. The new M4 motorway and the railway inevitably affect the peace and quiet of this stretch, although the country to the south of the Kennet improves steadily as it progresses westwards. Berkshire is well-known as orchard country. At Theale there is the first of the swing bridges that occur along the Kennet & Avon. Fortunately, since the completion of the M4, this bridge has reverted to carrying relatively infrequent road vehicles, so the passage of a boat no longer causes a major traffic hold-up. Opening the bridge is hard work, but there are instructions to help you. Make sure that you close the traffic barriers first of all and open them behind you. Theale village is ½ mile north of the bridge. After Theale the Kennet flows steadily through wooded fields towards Sulhamstead and reaches Tyle Mill after a pleasant tree-lined straight cut. The navigation now stops at Tyle Mill, where there is an attractive BWB mooring site, a slipway, a shop and a small car park for visitors. The moorings are administered by the BWB section inspector *(see below)*. The lock is currently (1973) being restored by the Kennet & Avon Canal Trust as part of the scheme to re-connect the Newbury cruiseway to the Thames at Reading.

Sulhamstead
Berks. PO, tel, stores. A scattered village ¼ mile SE of Tyle Mill, but with no real centre. There are several large houses standing in their own grounds; the most impressive is Folly Farm, built by Lutyens in 1906 in a William and Mary style. In 1912 Lutyens extended the house, this time using a Tudor style. The mixture of the two periods is most successful. The house is private.

Theale
Berks. Pop 1600. EC Wed. PO, tel, stores, garage, bank, station. ¾ mile NW of Shenfield lock. Although largely a Reading suburb, Theale has been given a new lease of life by the opening of the bypass and the M4 motorway. The main street is now quiet and relatively traffic-free, and the Georgian terraces can be enjoyed. The large church with its tall tower is interesting. It was designed by E. W. Garbett and built 1820–32 in a style based entirely on Salisbury Cathedral. Theale station is halfway between the town and Shenfield swing bridge.

BOATYARDS & BWB

BWB Burghfield bridge. (Enquiries to BWB Section Inspector, Lower wharf, Padworth, Reading. Woolhampton 2277). Moorings, water, refuse, sewage, BWB car park.

PUBS

Three Kings Jacks Booth Bath rd, Sulhamstead, ½ mile N of Tyle Mill.

Cunning Man Burghfield bridge. Canalside. Food (hot lunches).

Knight's Farm Restaurant Burghfield. (Reading 52366). ½ mile S of Burghfield bridge. Interesting variety of food prepared by Swiss chef. Good Food Guide 1973. *Closed Sun dinner and all Mon.*

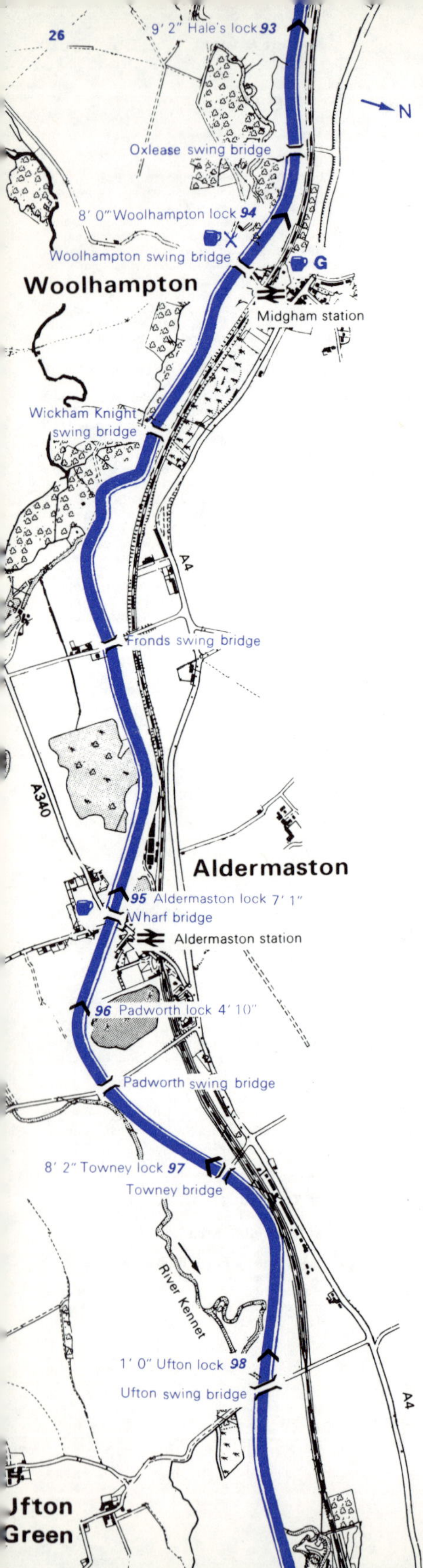

Woolhampton

5 miles

Leaving Tyle Mill, the present limit of navigation from Reading, the canal continues southwest, constantly joining and leaving the river Kennet. Swing bridges are very common, often carrying busy roads; these bridges are nearly all rusted up due to total disuse for several years. The towpath, which crosses constantly from one side to the other, is in good condition for walking – which is just as well, since walking is the best way to enjoy a canal whose locks are derelict and unusable. There is, however, a plentiful supply of water in the canal which makes it an excellent route for energetic canoeists (the Devizes-Westminster canoe race every Easter uses the Kennet & Avon canal for most of the way). The A4 runs parallel for many miles, but always keeps its distance; the Great Western Railway also runs parallel but much closer. There are attractive stations at Aldermaston and Woolhampton.

Woolhampton
Berks. PO, tel, stores, garage, station. A village on the A4 that owes its existence to the days of mail coaches on the old Bath road. There is a good mixture of buildings in the main street, several pubs and hotels, and, to the south, a charming GWR station, still resplendent in brown and cream. Up on the hill to the north of the village are the Victorian church, the Georgian buildings of Woolhampton Park and Douai abbey and school, the latter a fine group of 19thC buildings with more recent additions.

Aldermaston
Berks. PO, tel, stores. Attractively placed at the foot of a wooded hill, 1½ miles to the south of Aldermaston wharf (along a minor road), the village is particularly fine. Mellow brick houses of all periods face each other across the sloping main street, which has survived the inroads of traffic. At the top of the street is the pebble-dashed church, and Aldermaston Court, a private house containing magnificent 17thC woodwork. Nuclear weapons are no longer made at Aldermaston, but earthenware is; the pottery is in the main street, and can be visited.

Aldermaston Wharf
Berks. PO, tel, stores, station. A small canalside settlement bisected by the busy A340. The turf-sided lock, one of the oldest on the whole canal, is derelict, and the swing bridge is fixed, but the pub thrives. Perhaps this hamlet will one day rediscover its *raison d'être*.

Ufton Green
Berks. A lush peaceful hamlet built round a small triangular green. All that remains of the church is one flint wall, standing proudly in the middle of a field, and capped with a marvellous mantle of ivy.

BOATYARDS & BWB

BWB Padworth Yard near Aldermaston wharf. (Woolhampton 2277). The only BWB yard with no boats. Information about the canal, and moorings on the navigable sections. Special K & A-sized windlasses for sale.

PUBS

Row Barge Woolhampton. (2213). Canalside. Free house. Lunches daily, dinners at 12hrs notice.
Angel Hotel Woolhampton. (3307). Lunches, snacks and dinners. Also bed & breakfast.
Falmouth Arms Woolhampton.
Butt Inn Aldermaston wharf. Food.
Hinds Head Aldermaston. Food.

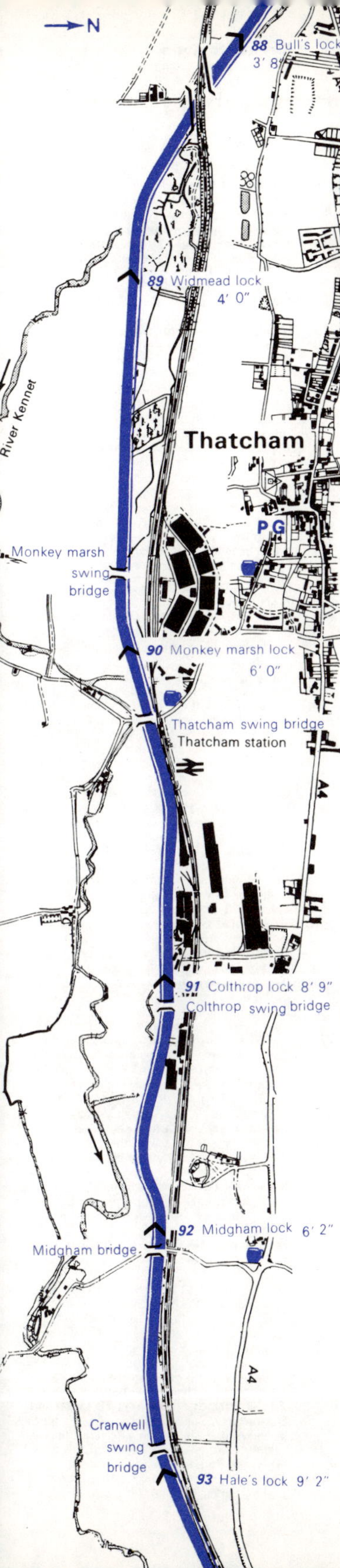

Thatcham

4½ miles

The unnavigable canal leaves Midgham Park to the north and continues due west through water meadows, the woods and hills receding to the south. At Colthrop a large industrial estate appears unexpectedly beside the canal; much of it consists of paper mills. Thatcham station is conveniently beside the canal; an hotel is nearby. The village itself is a mile to the northwest. The canal now flows very straight through isolated water meadows under a railway bridge to Bull's lock: beyond this derelict lock the Kennet & Avon becomes fully navigable for several miles, forming the Newbury cruiseway which now extends to Brunsden lock.

Thatcham
Berks. Pop 1,500 EC Wed, PO, tel, stores, garage, station. The main square of this expanding village is set back from the A4, and so it manages to retain some peace. The church is mostly Victorian, but at the end of the main street there is a 14thC chapel, now used as a school.

PUBS

Swan Hotel near Thatcham station.
Broadway Restaurant Thatcham.
Coach and Horses ½m N of Midgham bridge. Food.
Vane's 7 High Street, Thatcham. (63344). French provincial cooking, with particularly excellent puddings. Good Food Guide 1973. *Closed Sun & Mon.*

FISHING

At *Aldermaston* there are a few big trout as well as a variety of coarse fish. The chub fishing goes like a bomb (?) with fish to 6lb. Dace and grand roach fishing make for a fine day's sport.
Woolhampton offers excellent fishing for roach, chub and barbel; and there are sizeable dace, perch and pike.
At Midgham there are some attractive productive swims in the area. All the usual coarse fish, plus trout and occasional grayling. This area produced a trout of just under 11lb a few years ago, and a pike of 26lb has also been recorded.
Thatcham has some of the finest roach fishing on the Kennet: many specimen roach over 2lb have been caught. There are also good chub with the best reported at 7lb. Other species worth fishing for are dace, barbel and occasional tench.

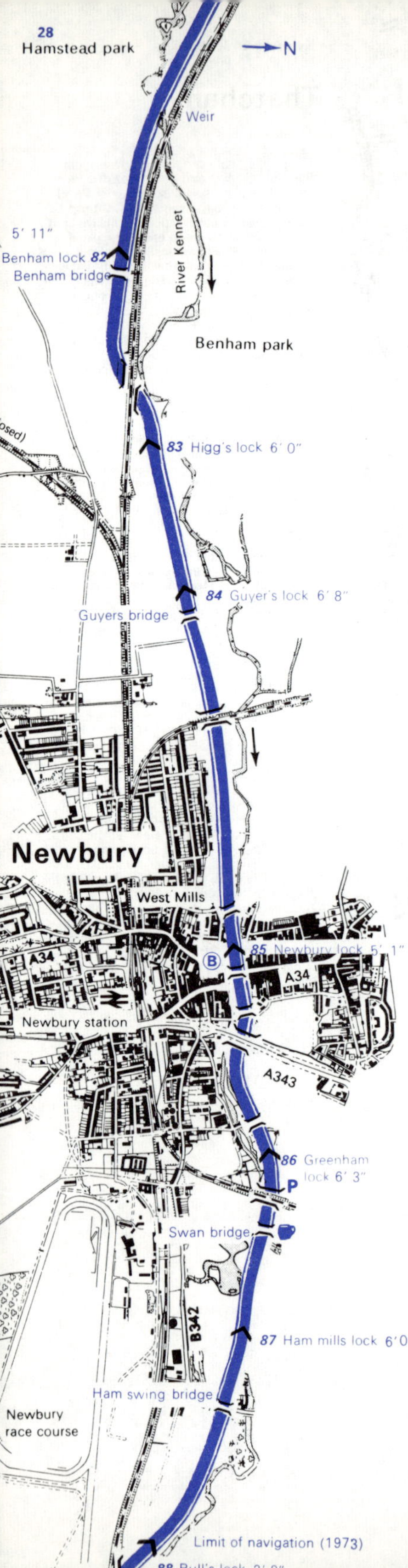

Newbury

4½ miles

Leaving Bull's lock, and Newbury race course, (2½m east of Newbury) this isolated navigable section of the canal/river passes Ham lock and enters Newbury, under a handsome new road bridge. Just beside this bridge is Newbury bus station, where vast acres of tarmac surround a decayed stone wharf building and a row of superb old warehouses which have been ingeniously converted into bus station buildings. This large wharf used to be the terminus of the Kennet Navigation from Reading, before the Kennet & Avon Canal Company extended it to link up with the Avon at Bath. West of the wharf the channel gets narrower and faster until it reaches a splendid stone balustraded bridge. Just beyond is Newbury lock, where there is a mooring site. The river cuts right through the town, and the town makes the most of it. West of the lock is the delightful, quiet West Mills area where rows of terraced houses face the navigation. There have been proposals to replace the old mill by an hotel. A narrowboat here does trips along the canal in summer. West of Newbury, the navigation again passes through extensive water-meadows before the wooded hills of Hamstead Park close in from the south.

Newbury
Berks. Pop 22,200. EC Wed, MD Tue. Newbury developed in the Middle Ages as a cloth town of considerable wealth, its stature indicated by the size of the church. Although the cloth trade has long vanished, the town has managed to retain much of its period charm. It is a busy shopping centre, and the shops fronts in the main streets have buried many 17thC and 18thC houses. Elsewhere in the town the 18thC is well in evidence, especially in the West Mills area. There are fine almshouses, and the pretty, ornamental stone bridge over the navigation. There are also signs of the agricultural importance of Newbury; the 19thC Italianate Corn Exchange, for example.

St Nicolas Church West Mills. Borders the canal on the south bank. A large perpendicular church, built *c.* 1500 at the height of Newbury's prosperity as a wool town. Its 17thC pulpit is most unusual.

St Nicolas School Enborne rd. By Butterfield, 1859.

Borough Museum Wharf rd. Originally built in 1626 as a cloth-weaving workshop to give employment to the poor, this is one of the most interesting buildings in Newbury. Adjoining is the corn store, once on the edge of the Kennet wharf. The museum collection illustrates the prehistoric and Saxon history of the region, as well as the medieval and modern. Also a natural history section with an excellent display of moths and butterflies. Models illustrate the Battles of Newbury. *Closed Sun, EC Wed.*

Round Barrow Cemetery Wash Common, near the site of the 1st Battle of Newbury in 1643. Memorial stones to the victims surmount the two smaller mounds.

Newbury Fair Northcroft lane, Northcroft. Leave canal at Kennet bridge. Annual Michaelmas fair held since 1215. *Thur following 11 Oct.*

1st Battle of Newbury, 20 Sept 1643
Site of Wash Farm off A343, 1¾ miles S of Guyers lock. Apart from Gloucester, Charles I held the west at the time of the battle. Both Royalists and Parliamentarians were marching towards London when battle was opened at Newbury. The Royalists were defeated by the Parliamentarians in one of the most bloody onslaughts of the Civil War. Guyer's and Higg's locks are named after troop commanders in the battle.

2nd Battle of Newbury, 28 Oct 1644
Donnington Castle, Donnington, 1½ miles N of Newbury lock off the A34. The Royalists were in possession of Donnington Castle when the Parliamentarians attacked under Skippon, Cromwell and Balfour. For one night Charles' army withdrew to Oxford,

but a week later they returned, relieved the castle and recovered their artillery. There is a reconstruction model of the battle in Newbury Museum.

BOATYARDS & BWB

Ⓑ **John Gould Waterways** Newbury lock, Newbury. (1548). Moorings, repairs (adjacent), water, slipway, refuse; sewage, day boats and houseboats for hire. Camp site for tents at Newbury lock.

BOAT TRIPS

Kennet Horse Boat Co. Horse-drawn boat trips from West Mills. (Newbury 4154). Available for private charter *except Suns and B. Hols*, when a public 3-hour trip is run at *15.00*. Licensed bar and teashop on board.

PUBS

☕✕ **Queen's Hotel** Market place, Newbury. (47). Restaurant.
✕🍷 **La Riviera** 26 The Broadway, Newbury. (2074). Greek restaurant: in the Good Food Guide since 1963
✕ **Tudor Cafe** by Newbury bridge. Patisserie and boulangerie.
☕✕ **White House** by Swan bridge, Newbury. Food.

CARAVANS & CAMPING

Elmhurst Caravan Park Elmhurst rd, Thatcham. $1\frac{1}{4}$ miles NE of Ham Mills lock. (3114). Short stays allowed, although hire only for long lets.
Sandford Lodge Caravan Site Winchester rd, Newbury. 1 mile S of Newbury lock. (Boxford 287). Tents and caravans for hire.

FISHING

Newbury has gained a reputation for dace fishing, with many specimens over a pound. Other species worth fishing for are roach and chub, also trout and grayling. Several trout of 5 to 6lb have been caught around here.

Newbury.

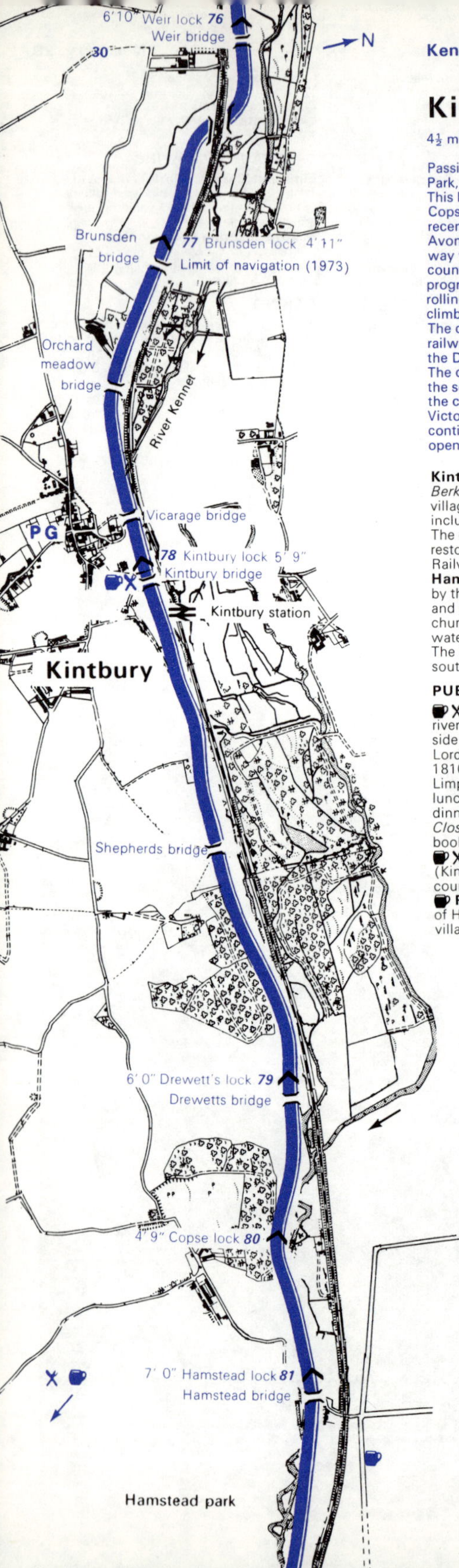

Kintbury

4½ miles

Passing the beautiful woods of Hamstead Park, the canal reaches Hamstead lock. This lock, and the three that follow it, Copse, Drewett's and Kintbury have recently been restored by the Kennet & Avon Trust, extending the Newbury cruise-way westwards towards Hungerford. The country opened up to boats by this programme is particularly attractive. Wooded rolling hills flank the canal to the south as it climbs up the locks towards Kintbury. The canal enters the village beside the railway, passing the splendid station, and the Dundas Arms, which overlooks the lock. The centre of Kintbury is up on the hill to the south of the lock. Leaving the wharf, the canal follows the railway, passing the Victorian Gothic vicarage, and then continues westwards through pleasing open countryside.

Kintbury
Berks. PO, tel, stores, station. A quiet village with attractive buildings by the canal, including a watermill and canalside pub. The church is originally 13thC but was restored in 1859; the Great Western Railway lends excitement to the situation.
Hamstead Park A very fine park bordered by the canal. There used to be a castle here and several interesting buildings adjoin the church on the side of the hill. There is an old watermill by the newly-restored lock. The hamlet of Hamstead Marshall lies to the south, 1½ miles from Hamstead lock.

PUBS

Dundas Arms Kintbury. (263). The river Kennet and the canal flow on either side of this pub, which was named after the Lord Dundas who opened the canal in 1810, giving his name to the aqueduct at Limpley Stoke. The food is good; simple lunches contrast with more elaborate dinners, including of course Kennet trout. *Closed all Mon, and Sun evening.* Must book. In the Good Food Guide since 1963.
White Hart Hamstead Marshall. (Kintbury 201). Restaurant in an old country pub. 1 mile S of Hamstead lock.
Red House Marsh Benham. ¼ mile NE of Hamstead lock. Pub in a thatched estate village near Benham Park.

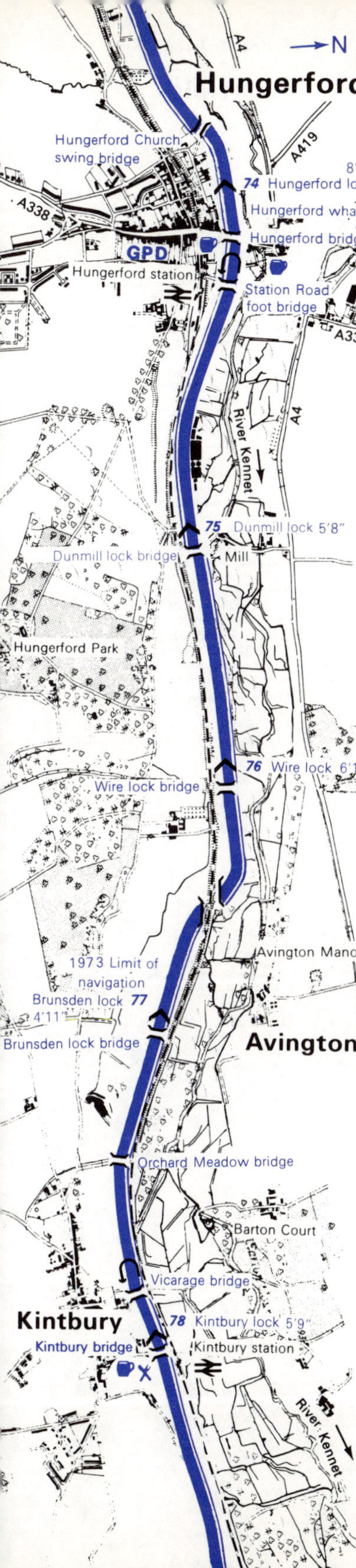

Hungerford

4½ miles

The canal continues westwards through open countryside. The railway and the river Kennet are constantly present, the river leaving the canal for the last time west of Kintbury. Locks 77 and 76 carry the canal past Avington, with its Norman church visible among the trees. Pretty woods accompany the canal to the south as it approaches Hungerford, while to the north river and canal run side by side through water meadows, separated only by a narrow ridge carrying the towpath. By lock 75 the towpath turns over to the south bank. From the bridge there is a good view of Denford Mill. As the canal enters Hungerford, the Kennet swings away to the north, feeding the trout farm that lies between canal and river. Gardens flank the canal as it comes into the centre of the town. Access to Hungerford is easy by the town bridge, which leads directly to the handsome main street.

Hungerford
Berks. Pop 3,700. EC Thur. PO, tel, stores, garage, bank, station. Hungerford is built along the A338 which runs through the town, southwards from the junction with A4. The pleasant 18thC and 19thC buildings are set back from the road, giving the spacious feeling of a traditional market town. None of the buildings is remarkable, but many are individually pretty. Note the decorative ironwork of the house by the canal bridge. The manor was given to John of Gaunt in 1366, and any monarch passing through the town is given a red rose, the Lancastrian emblem, as a token rent.

Hocktide Ceremonies
On the second Tuesday after Easter, 99 commoners (those living within the original borough who have the rights of the common and the fishing) are called to the Town Hall by the blowing of a horn. Two Tuttimen are appointed who have to visit the houses of the commoners to collect a 'head penny' from the men, and a kiss from the women: they give oranges in return. All new commoners are then shod by having a nail driven into their shoes. This ceremony dates from the medieval period.

Avington
Berks. The village is best approached along the track that runs east from lock 76, although the more adventurous can go directly across the water meadows, crossing the Kennet on a small footbridge. The little church is still wholly Norman, and contains a variety of original work; the chancel arch, the corbels and the font are particularly interesting.

PUBS

John of Gaunt Hungerford. (2642). Food. B&B.

Three Swans Hungerford. (2721). Food. B&B.

Thompson's Eating House 17 High st, Hungerford. (2056). Victorian-style restaurant serving French and Swiss food. Home-made ice cream. Good Food Guide 1973. Closed Sun, 2 weeks in Mar.

FISHING

Specimen roach may be found in the Hungerford area; many examples over 2lb have been caught. There are also chub, dace, tench and pike.

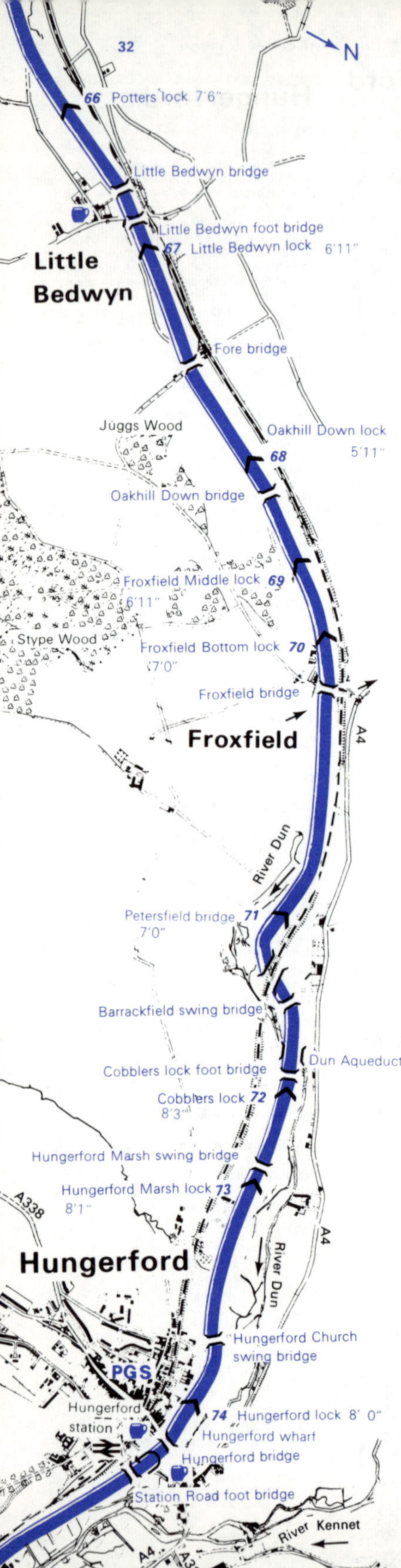

Kennet & Avon

Froxfield

4¾ miles

Leaving Hungerford, the canal passes the old wharf. An original stone warehouse survives, but much of the wharf area has now been built on. West of the wooded 19thC church, with its pleasantly overgrown churchyard, the canal suddenly enters an open landscape. Water meadows and pasture land, rich in buttercups, flank the waterway which seems to be more river than canal. The railway is in a cutting to the south, and so the quiet and the browsing cattle give a feeling of 18thC rural serenity. Even the broken-down locks add to this feeling, seemingly part of a Constable landscape. The canal is then carried on to an embankment as the wooded hills reappear on both banks, crossing the river Dun on a small brick aqueduct. The railway crosses the canal west of the aqueduct, and now hugs the north bank for several miles. The roar of the frequent diesel expresses to and from the west country is the only interruption in the natural peace and solitude of the canal. Froxfield lies to the north, flanking the A4; the best access is from the new bridge. This was rebuilt in 1972 during a road improvement scheme; surprisingly it was rebuilt in the old style, using traditional methods and materials, even to the correct colour of brick. Three locks carry the canal past Froxfield, and then the spire of Little Bedwyn church comes into view, half hidden by trees on the north bank. The village is cut in half by the canal and the railway. In the centre the lock continues the climb towards the summit.

Little Bedwyn
Wilts. PO, tel. Divided by the canal, the village falls into two distinct parts. North is the estate village, pretty 19thC terraces of patterned brick running eastwards to the church, half hidden among ancient yew trees. To the south is the older farming village, handsome 18thC buildings climbing the hill away from the canal.

Froxfield
Wilts. PO, tel, stores. The village is ranged along the A4, which has obviously affected its development. Some attractive houses have survived the onslaught of the A4 traffic. The main feature of the village is the Somerset Hospital, a range of almshouses founded by the Duchess of Somerset in 1694, extended in 1775 and again in 1813. Facing on to the road, the Hospital is built round a courtyard which is entered by a Gothic-style gateway, part of the 1813 extension.

Littlecote
Wilts. 1½ miles N of Froxfield. A Tudor building of the 16thC, Littlecote is the most important brick mansion in Wiltshire. The formal front overlooks the gardens that run down to the Kennet. Inside the Great Hall, the armoury and Long Gallery are particularly notable. *Open: Apr-mid Oct. Tue, Wed, Sat 14.00–17.00, (Sun 14.00–18.00.) Winter opening by appointment.* (Hungerford 2509).

PUBS

Harrow Little Bedwyn. In the southern half of the village.

Pelican Froxfield. On A4, ¼ mile N of the new bridge.

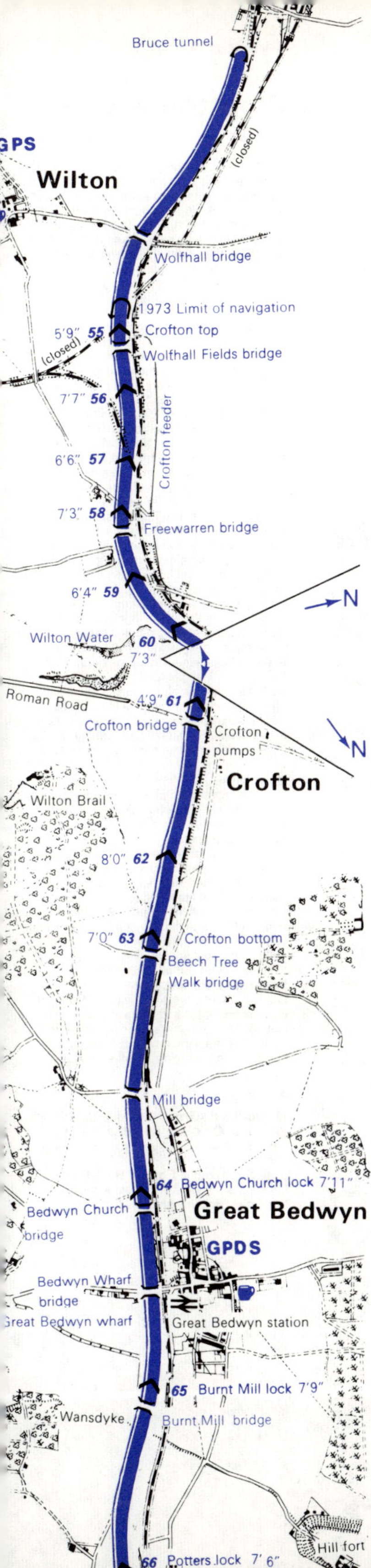

Crofton

$4\frac{1}{2}$ miles

Leaving Little Bedwyn, the canal continues through a rolling landscape towards its summit, closely accompanied by the railway. To the north is a hill fort, overlooking ridges that break up the farmland. The canal stays on the south side of the valley, a shallow side cutting carrying it into Great Bedwyn. The village is ranged over the hillside to the north of the canal, newer houses spilling downwards towards the canal and railway station. The canal leaves the village past the church, and enters a wooded stretch that takes it up towards Crofton. The hills encroach more sharply as the summit draws nearer. Crofton appears as the canal starts a wide swing to the north west. The engine house stands on a rise above the canal, its blunt, iron-bound chimney making its purpose unmistakeable. To the south lies the long expanse of Wilton Water, a natural lake from which the Crofton pumps draw their supplies. After Crofton the country opens out for a while as the flight of locks continue the final climb to the summit. Parts of the canal are dry and overgrown west of Crofton, but water reappears by lock 57 as the canal passes the site of the old railway bridge. Two more locks bring the canal to its summit level. The canal immediately passes the site of another rail bridge, and then as the land rises steeply on both banks, it prepares itself for the short Bruce Tunnel. A wooded cutting leads towards the tunnel, taking the canal through the fringes of the old Savernake Forest. To the north are the extensive parklands of Tottenham House, and Savernake Forest itself. The towpath climbs over the top of the tunnel.

Wilton
Wilts. PO, tel, stores. A compact village at the southern end of Wilton Water, with a pretty duck pond in the centre.

Wiltshire Wildlife Park
$\frac{1}{2}$ mile E of Wilton, accessible from the road crossing the tail of lock 61. Family orientated nature reserve, with children's boating pond, pets' corner, picnic sites. No dogs. *Open: Nov-Mar, Sats & Suns only. Mar-Nov daily.*

Crofton
Wilts. The scattered village is dominated by the brick pumping house with its separate chimney. It houses two 19thC steam engines, one built in 1812 by Boulton and Watt, the oldest working beam engine in the world, the other in 1845 by Harveys of Hayle, Cornwall. Both have been restored to working order, and are steamed on several weekends in the year. The pumping house and the engines are open for viewing *every Sunday, 10.00–18.00.* For details of 'steaming' weekends, write to: Crofton Beam Engines, 11 The Vineyard, Richmond, Surrey.

Great Bedwyn
Wilts. PO, tel, stores, garage, station. The main street climbs gently away from the canal and the railway. It is wide, with generous grass verges; attractive houses of all periods line the street. At the top is the pub. The large church, with its well-balanced crossing tower, is mostly 12thC and 13thC; inside are some interesting monuments. The road running westwards to the church passes the Bedwyn Stone Museum, an amazing establishment.

Bedwyn Stone Museum
A collection of stone work of all types, showing the work of seven generations of stone masons. There are statues, tombstones, casts, even the fossilised footprint of a dinosaur. Open regularly, details from B. J. Lloyd. (Great Bedwyn 234).

PUBS

Swan Inn Wilton, $\frac{1}{2}$ mile S of lock 61, on road running beside Wilton Water.

Three Tuns Great Bedwyn.

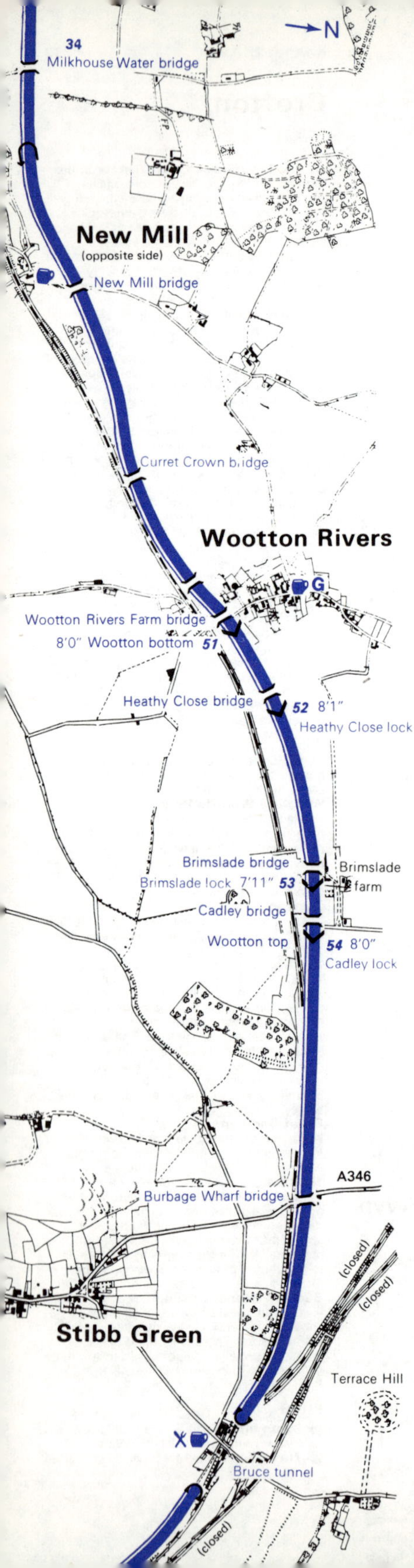

Milkhouse Water bridge
N
New Mill
(opposite side)
New Mill bridge
Curret Crown bridge
Wootton Rivers
G
Wootton Rivers Farm bridge
8'0" Wootton bottom 51
Heathy Close bridge
52 8'1"
Heathy Close lock
Brimslade bridge
Brimslade
Brimslade lock 7'11" 53
farm
Cadley bridge
Wootton top
54 8'0"
Cadley lock
A346
Burbage Wharf bridge
(closed)
(closed)
Stibb Green
Terrace Hill
Bruce tunnel
(closed)

Wootton Rivers

4¾ miles

The canal emerges from the western portal of Bruce Tunnel into a deep cutting. Woods line both banks, hiding the railway which is now on the south bank, having crossed over the tunnel. The towpath passes under the railway and descends steeply to the canal. The cutting continues westwards to the high brick bridge that carries the A346, and then the landscape opens out: the rolling hills still follow the canal, but recede slightly. Immediately after the bridge is Burbage wharf; several of the original brick canal buildings still stand, attractively converted to domestic use, and the old wooden wharf crane still hangs over the water. Pasture and arable land flank the canal on its course to the first of the four Wootton locks. This flight ends the short summit, and starts the long descent towards Bath. By the first lock there is a pretty cottage and garden, while the second is in the middle of Brimslade farm, whose attractive tile-hung buildings date from the 17thC. The last two locks take the canal to Wootton Rivers; the houses stretch northwards away from the canal, which is overlooked by the church. The Wootton Rivers flight is being restored at present, and should be back in use by 1973; this will extend the navigation to Crofton top lock, adding the three mile summit level to the 15-mile-long Devizes pound. As the country undulates, the canal maintains its level, moving alternately from low cutting to low embankment. Woods break up the hills, giving fine views to the south. New Mill is a small hamlet south of the canal with a convenient pub; there is also a small wharf. At the end of this section the railway moves away to the south to pass through Pewsey, leaving the canal in peace at last.

New Mill
Wilts. A pretty hamlet scattered below the canal. The mill which gave it its name is now a house, with a fine garden.

Wootton Rivers
Wilts. PO, tel, stores. A particularly pretty village composed almost entirely of timber-framed thatched houses, climbing gently up the hill away from the canal. Even the walls by the canal are thatched. All the houses are attractive, including the large manor by the canal. The little church with its wooden bell turret is set among trees; it was extensively rebuilt in the 19thC.

Savernake Forest
Wilts. A small village grew up in the 19thC around the hotel and the two railway stations, to cater for an early holiday trade. Today the stations have vanished, but the hotel still thrives, on the hill above Bruce Tunnel.

PUBS

Liddiard Arms New Mill. Bar billiards.

Royal Oak Wootton Rivers. Bar billiards. Very attractive pub in the main street.

Savernake Forest Hotel Savernake Forest. (Burbage 206). Restaurant. The hotel has fishing rights on the canal.

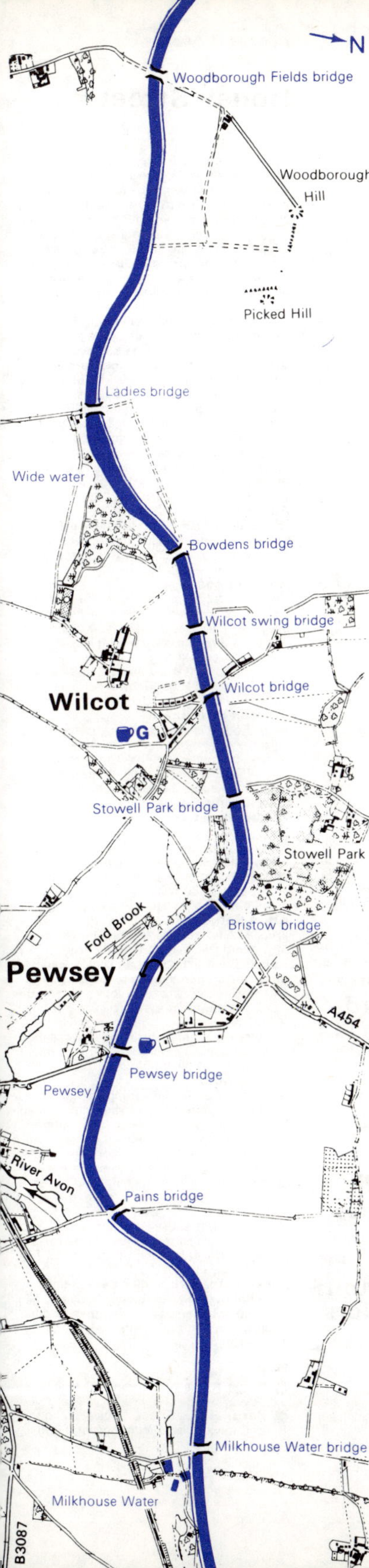

Pewsey

4¾ miles

The 15-mile-long pound continues westwards towards Devizes through rolling hills, forming a landlocked stretch of navigable waterway in the middle of the canal. To the north, hills descend to the water's edge, and to the south the land opens out, giving fine views over the Vale of Pewsey. The canal crosses this landscape with a mixture of cutting and embankment, passing through woods from time to time. The railway lies to the south, and is now out of sight. The canal turns towards Pewsey, but still passes well outside the town, which fills the Vale to the south. Pewsey wharf is ½ mile from the town centre, and so has developed as a separate canalside settlement, with a pub, cottages, and warehouse buildings. The canal leaves Pewsey in a low wooded cutting, swinging back to its usual westerly course. The woods continue past Stowell Park, whose landscaped grounds extend to the north. The house, built early in the 19thC, can clearly be seen from the canal. A miniature suspension bridge carries a private footpath from the Park across the canal. A straight stretch leads to the first cottages of Wilcot; the rest of the village is to the south. The steep bare mound of Picked Hill dominates the canal as it passes Wilcot and enters the wooded Wide Water, a natural pond incorporated into the canal. This is terminated by the unusual Ladies bridge: dated 1808, and attributed to Rennie, this decorative stone bridge is rich in neo-classical ornament. It is a unique bridge on the canal, for such splendour was usually reserved for major engineering works. The canal skirts Picked Hill, giving a good view of the field terracing that is a relic of Celtic and later mediaeval cultivation. The equally dominant Woodborough Hill now fills the north bank, while to the south open country leads to the village of Woodborough.

Wilcot
Wilts. PO, tel, stores. A pretty village scattered round the green; there are several thatched houses, a little village school with a prominent bell, and a blacksmith. Parts of the church date from the 12thC, but it was mostly rebuilt in 1876 after a fire.

Pewsey
Wilts. Pop 2,500. EC Wed. PO, tel, stores, garage, bank, station (but very few trains stop). The little town is set compactly in the Vale of Pewsey. At its centre, overlooking the young river Avon is a fine statue of King Alfred, erected in 1911. From this all the roads radiate. There is the usual mixture of buildings; but while many are attractive, none is noteworthy. The church is mostly 13thC and 15thC, but parts of the nave are late Norman: the altar rails were made from timbers of the 'San Josef' captured by Nelson in 1797.

Pewsey White Horse
1½ miles S of the town. Dating from the 18thC, the horse was re-cut in 1937 to celebrate the coronation of George VI. It is 66ft long.

BOAT TRIPS

The paddle-boat 'Charlotte Dundas' operates trips along the 'long pound' from Pewsey Wharf to Devizes. See page 38.

PUBS

Golden Swan Wilcot. Food, B&B. At the far end of the village, overlooking the green.

French Horn Pewsey. Canalside, on A345 by wharf.

Royal Oak Pewsey. (3426). In town centre. Food, B&B.

Phoenix Pewsey. (2458). In town centre. B&B.

Horton
N
Horton Chain bridge
Etchilhampton Water
Horton Fields swing bridge
The Knoll
Allington bridge
Allington
Allington swing bridge
Woodway bridge
G
All Cannings
All Cannings bridge
West Stanton bridge
Stanton
St Bernard
Stanton bridge
White Horse →
Alton →
Barnes
Honey Street
Honey Street
wharf
PG
Honey Street bridge
Alton
Priors
Ridge Way →
Alton Valley bridge
Woodborough Fields bridge

Honey Street

$5\frac{1}{4}$ miles

Leaving Woodborough Hill behind, the long pound continues westwards towards Devizes. To the south the land falls away, while to the north the tower of Alton Priors church comes into view. Beyond the village can be seen the white horse cut into the hill in 1812, a copy of the one at Cherhill. The canal passes Honey Street wharf, one of the best preserved on the canal, with a fine collection of original buildings, and a canalside pub. The canal now begins to meander through the open countryside, roughly following a contour line to maintain its level. Its progress is marked by a succession of shallow cuttings and low embankments. Several villages are near the canal, all visible and easily accessible from the many bridges, but none actually approach the waterside. Their interests lie rather in the rich agricultural lands that flank the canal. Leaving Allington the canal curves round the Knoll, a major feature of the landscape to the north.

Allington
Wilts. Tel. A small agricultural village scattered round the Victorian church. East of the village is All Cannings Cross, a large Iron Age settlement.

All Cannings
Wilts. PO, tel, stores. An attractive village built round a square, with houses of all periods. To the south there is a large green, overlooked by the church with its tall central tower. Although the church is mostly 14thC, its most interesting feature is the ornamental High Victorian chancel, added in 1867.

Stanton St Bernard
Wilts. PO, tel, stores. Built in a curve of the hills, the village has one main street, flanked by pretty gardens. The best building is the 19thC manor, which incorporates relics of an earlier house. The battlemented church is Victorian.

Honey Street
Wilts. A traditional canalside village. There are brick cottages, a warehouse, and some pretty weatherboarded buildings. The wharf of course cannot be used as such, but seems only to be waiting for the canal to come back to life.

Alton Barnes
Wilts. PO, tel, stores, garage. The village runs along the road northwards from Honey Street. The best part is clustered round the church. Fine farm buildings and an 18thC rectory are half hidden among the trees. The church is essentially Anglo-Saxon, but has been heavily restored; everything is in miniature, the tiny gallery, pulpit, and pews emphasising the compact scale of the whole building.

Alton Priors
Wilts. Tel. Approached along a footpath from Alton Barnes churchyard; the isolated church is the best feature of this scattered hamlet. This pretty Perpendicular building with its wide, well-lit nave contains a most interesting monument; a big box tomb is surmounted with a large engraved Dutch brass plate, dated 1590, rich in extravagant symbolism. To the east of the village the Ridgeway runs southwards towards Salisbury. This Bronze Age drover's road swings north-east along the downs for 50 miles, finally joining the Thames valley at Streatley. The path is not clearly defined here, but five miles to the north, by the A4 crossing, it becomes a wide unmistakeable track which continues unbroken to the Thames.

PUBS

Kings Arms All Cannings. $\frac{1}{4}$ mile S of Woodway bridge.

Barge Honey Street. Canalside. Food. B&B. Caravan hire. (Woodborough 238).

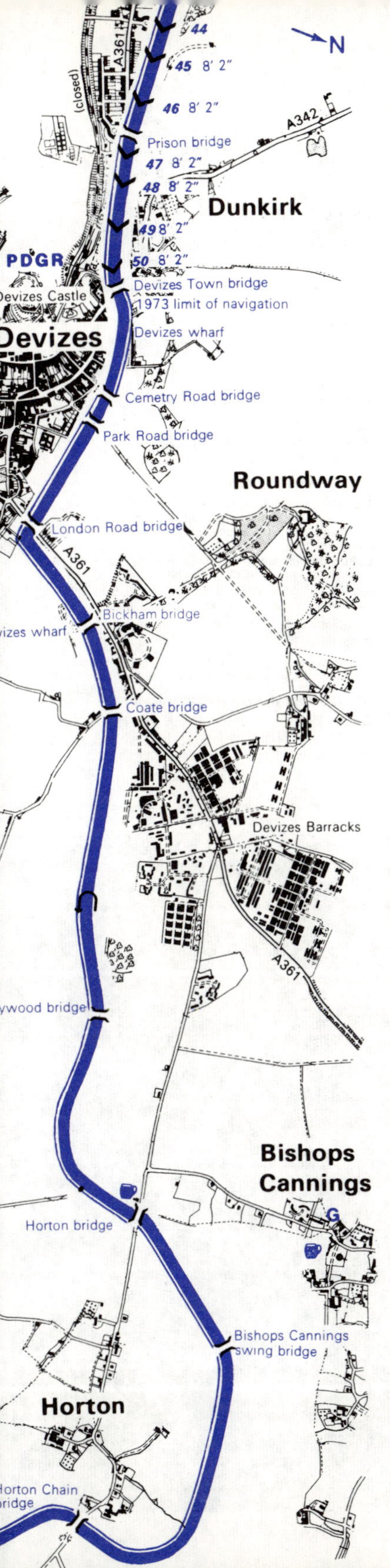

Devizes

$5\frac{1}{4}$ miles

Leaving Horton, the long pound continues westwards towards Devizes. Following the contour of the land, it swings in a wide arc towards Bishops Cannings. The rolling hills climb fairly steeply to the north, while the pasture land falls away to the south. After a low cutting, the tower of Bishops Cannings church comes into view, half hidden by trees; a footpath from the swing bridge, now restored like all the others on the long pound, is the quickest way to the village. At Horton bridge, where there is a convenient canalside pub, the canal enters another short cutting. The landscape opens out again, to allow a view of the handsome Victorian barracks outside Devizes. The canal passes the barrack buildings, which are partly hidden by trees, and then enters the long wooded cutting that carries it through Devizes. Houses appear, their gardens overlooking the cutting, and the traffic noise on the busy A361 marks the return to civilisation. Several very elegant large stone bridges (many listed as ancient monuments) span the cutting. Access to the town is easy at all the bridges. At Cemetery Road bridge the towpath turns over to the north bank for a short stretch, returning to the south at the next bridge. Between these two bridges is the disused Devizes wharf, showing few signs of its former glory. Unfortunately Devizes has not yet taken advantage of the canal's presence. Beyond the wharf, and the gas works that follows it, the long pound ends at the first lock of the famous Caen Hill flight, preceded by the generous stone bridge with its separate towpath archway. Locks now come at regular intervals, preparing the canal for the dramatic descent down Caen Hill.

Devizes
Wilts. Pop 9,000. EC Wed. MD Thur. PO, tel, stores, garage, bank, cinema. Despite the effects of traffic, Devizes still retains the atmosphere of an old country market town. Originally the town grew up around the castle, but as this lost its significance the large marketplace became the focal point. Handsome 18thC buildings now command the square, while the market cross records the sad story of Ruth Pierce. Elsewhere there are timbered buildings from the 16thC. The two fine churches, one built for the castle and the other for the parish, tend to dominate the town, and hold it well together. Only the mound and related earthworks survive of the original Norman castle; the present building is an extravagant Victorian folly.

St. John's Church
Built by Bishop Roger of Sarum, who was also responsible for the castle, this 12thC church with its massive crossing tower is still largely original. There are 15thC and 19thC additions, but they do not affect the Norman feeling of the whole.

St. Mary's Church
Dating from the same time as St John's, this church was more extensively rebuilt in the 15thC; plenty of Norman work still survives, however.

Wiltshire Archaeological and Natural History Museum Long street. The collections include finds from Neolithic, Bronze and Iron Age sites in Wiltshire, the most famous being the Stourhead collection of relics excavated from burial mounds on Salisbury Plain. There are also Roman exhibits. *Open daily 10.00–17.00 during summer, 10.00–16.00 the rest of the year.*

Battle of Roundway Down, 13 July 1643 During the early months of the Civil War the Royalist army was particularly successful in the west. Devizes was held by a Royalist army that had already tested the Roundhead forces, who under General Waller were tired, dispirited and short of supplies after their defeat at Lansdown Hill, near Bath. Trying to recover lost ground, Waller drew up his forces on Roundway Down, near Devizes. A Royalist cavalry charge took Waller by surprise, and drove his army down the steep hill in disarray. At a given signal, the Royalist infantry sallied forth from Devizes and fell upon the confused and battle weary Roundheads. Most of Waller's army were killed or captured. This sweeping Royalist victory paved the way for the capture of Bristol on 26 July by Prince Rupert, which represented the high point of Royalist fortunes. The battlefield, off the A361 north-east of Devizes, is still largely intact, and can easily be explored on foot.

Coate
Wilts. PO, tel. A nondescript farming village built round a square; there is an interesting pub, the New Inn.

Wiltshire Regiment Museum Le Marchant Barracks, 1½ miles east of Devizes on A361; also accessible from the canal. The history of the regiment from its foundation in 1756 to the present day. *Open weekdays 10.00–16.30.*

Devizes to Westminster Canoe Race
The toughest and longest canoe race in the world takes place every Easter. The course, from Park Road bridge, Devizes, to County Hall Steps, Westminster, includes 54 miles of the Kennet & Avon, and 71 miles of the Thames, the last 17 of which are tidal. There are 77 locks. The race grew from a background of local rivalry in Pewsey and Devizes to find the quickest way to the sea by boat; in 1948 the target was 100 hours. In 1950 the first regular annual race over the course took place; three years later the junior class was introduced. Course times dropped rapidly, and by 1965 the record stood at 20 hours 27 minutes. The number of entries increases every year, and is now well above 300. Anyone may enter for the race, but they would have difficulty in beating the highly-trained army and navy teams from Britain and Europe.

Bishops Cannings
Wilts. PO, tel, stores. Apart from one or two old cottages, the main feature of this village is the very grand church. This cruciform building, with its central tower and spire, is almost entirely Early English in style; its magnificence is unexpected in so small a village. Traces of the earlier Norman building survive. Inside is a 17thC penitential seat, surmounted by a giant hand painted on the wall with suitable inscriptions about sin and death.

BOAT TRIPS

Charlotte Dundas
This paddle-driven narrow boat, operated by the Kennet & Avon Canal Trust, runs pleasure trips on the long pound, between Devizes and Pewsey. Details from: Bookings Manager, 2 Small st, Chirton, Devizes, Wilts. (Chirton 673).

PUBS

Bear Market Place, Devizes. (2444). This 18thC coaching inn has a French and English restaurant. Many other pubs and restaurants in Devizes.
New Inn Coate. Beer garden, with a small aviary of exotic birds.
Crown Bishops Cannings.
Bridge Inn Horton. Canalside, by Horton bridge. B&B.

CARAVANS & CAMPING

The Wharf Devizes. (3628). Centre of town adjoining public car park. Overnight stops approved for touring caravans and motor caravans. *Open all year.*
Lakeside Rowde, Devizes. (2767). 1 mile NW of Devizes on A342. Overnight stops approved for touring caravans, motor caravans and tents. Caravans for hire, gas, showers (h&c), shop. *Open Easter-Oct.*

FISHING

The Devizes area is good for big tench, apart from the usual selection of coarse fish.

TOURIST INFORMATION CENTRE

Devizes Northgate House, Northgate st, Devizes. (2160).

Devizes: Caen Hill locks in 1947, shortly before they fell into disuse. *Aerofilms Ltd.*

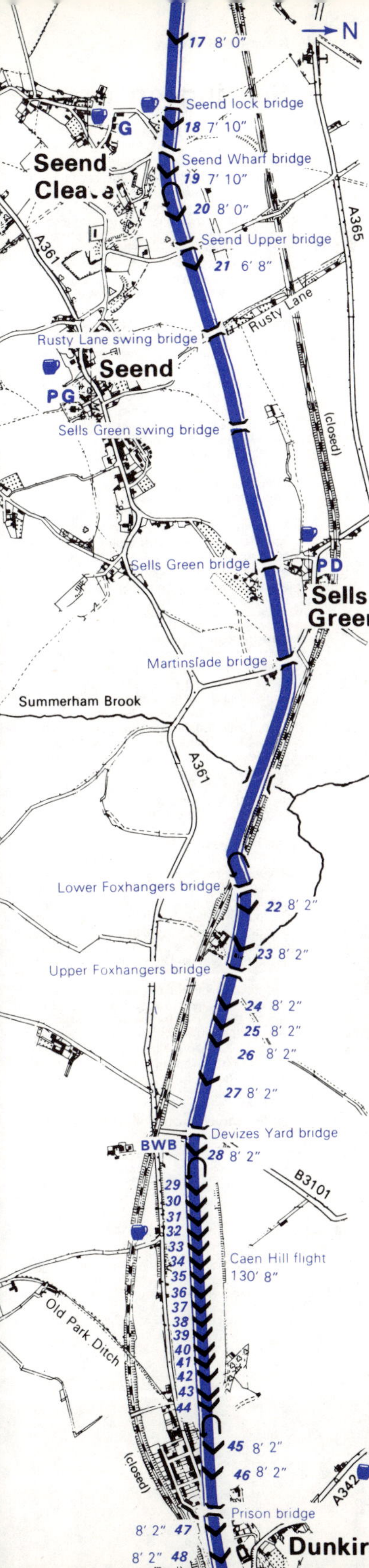

Sells Green

4½ miles

Leaving Devizes, the canal continues westwards. Ahead the landscape falls away, giving advance warning of the steep descent at Caen Hill. Locks occur at regular intervals west of Devizes, each separated by a long, wide pound. These were designed to hold sufficient water while permitting the locks to be close together to follow the slope. The locks are all derelict, but many of the pounds are in good condition and are full of water. The towpath is in very good condition; apart from the attraction to visitors of the Caen Hill flight, the whole area is obviously used for recreation by the people of Devizes. To the south the busy A361 accompanies the canal down the hill, but it is out of sight for most of the way. At lock 44 the major flight itself starts; wide lock follows wide lock down the hill, each with an enormous side pound. The scale of the whole flight is most impressive. All is of course unusable; the gates are missing, the intervening pounds silted or overgrown. One or two have even been turned into small enclosed pastures, horses now grazing where the water to fill the locks was stored. The task of restoration is enormous, but work is slowly getting under way. Pounds are being cleared, but the major task of lock restoration still lies ahead. Looking at the state of the flight today, it is difficult to imagine boats once again working their way up and down. At lock 29 the Caen Hill flight ends, but the locks continue the descent, now separated once again by longer pounds; like those at the top, these still hold water. The canal passes under the B3101 road bridge, and then at lock 22 reaches the end of the long fall—29 locks in two miles. By this last lock the towpath turns over to the north bank, and the canal is joined by the old railway, which follows it closely. The railway bridge has been removed. Water meadows accompany the canal to the south, with views of the hills that rise in the distance. The canal turns past Sells Green in a low cutting that hides most of the village, and then strides along the valley. The hills to the south climb steeply up to the village of Seend, and to the north flat pastureland stretches away. After two swing bridges the canal reaches the first of the five Seend locks; this is the best point for access to the village. By the third lock there is a pub, and a lane leading to Seend Cleave village.

Seend Cleave
Wilts. Tel, stores. An agricultural village built on the steep slopes of the hills that overlook the canal. Some new development has merged well with the existing houses.

Seend
Wilts. PO, tel, stores, garage. Although the main road cuts the village in half, Seend is still attractive. Elegant 18thC houses flank the road, and conceal the lane that leads to the battlemented Perpendicular church.

Sells Green
Wilts. PO, tel, garage. A scattered main road village, the houses doing their best to hide from the traffic behind decorative gardens.

PUBS

Brewery Inn Seend Cleave. ¼ mile from lock 19.

Bell Inn Seend. On A361. ½ mile from lock 21.

Barge Inn Seend. Canalside, by lock 19. Bar billiards. PO box.

Three Magpies Sells Green. On A365, ¼ mile from Martinslade bridge.

Olive Branch Inn Devizes. On A361 beside Caen Hill flight.

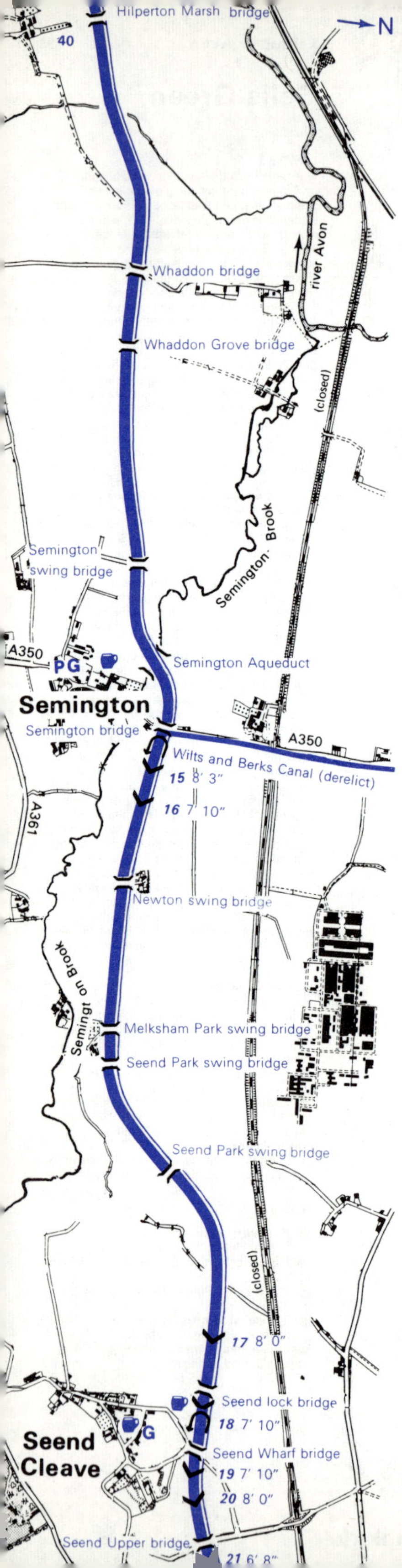

Semington

4½ miles

Leaving Seend locks behind, the canal continues its western course, maintaining a fairly straight line through open country. The steep hills around Seend are left behind, giving way to rolling farmland extending into the distance on both banks. This is a quiet and secluded stretch with no villages beside the canal; however farms occur regularly along the bank, each generally with its own swing accommodation bridge. In the distance to the north the embankment of the disused railway follows the course of the canal. The two Semington locks continue the descent towards Bath; the attractive lock house by lock 15 is being restored. Just beyond the lock the canal is crossed by the A350; this is the best access point for Semington. A close examination of the north bank just before the bridge will reveal a bricked-up side bridge; this marks the site of the junction with the long abandoned Wiltshire & Berkshire Canal, which used to go to Abingdon. Beyond the bridge the canal curves round past Semington on an embankment, crossing the Semington Brook on a small stone aqueduct. A long straight stretch now leads towards Hilperton, while to the south the hills return to follow the canal. To the north the river Avon draws gradually nearer, and the two waterways begin to share the same valley.

Semington
Wilts. PO, tel, stores, garage. Despite the main road, Semington is a pretty village. Large handsome houses with fine gardens run beside the road. Several date from the 18thC. The little stone church, crowned with a bellcote, is at the end of a lane to the east of the village. The old village school is beside the church, built in the same style.

The Wiltshire & Berkshire Canal
Opened in 1810, the canal wound in a meandering course for 51 miles between Semington on the Kennet & Avon and Abingdon on the river Thames. A branch was opened in 1819 from Swindon to connect with Latton on the Thames & Severn Canal. Although the carriage of Somerset coal was the inspiration for the canal, its eventual role was agricultural. Profits were never high, partly because the wandering line of the canal and its 45 locks made travel very slow, and so it suffered early from railway competition. By the 1870s moves were afoot to close the canal, and despite various efforts to give it a new lease of life, the situation had become hopeless by the turn of the century. Traffic finally stopped in 1906, and the canal was formally abandoned in 1914. Little remains of it now, but its course can still be traced with difficulty.

PUBS

Seymour Arms Semington. ¼ mile S of Semington bridge.

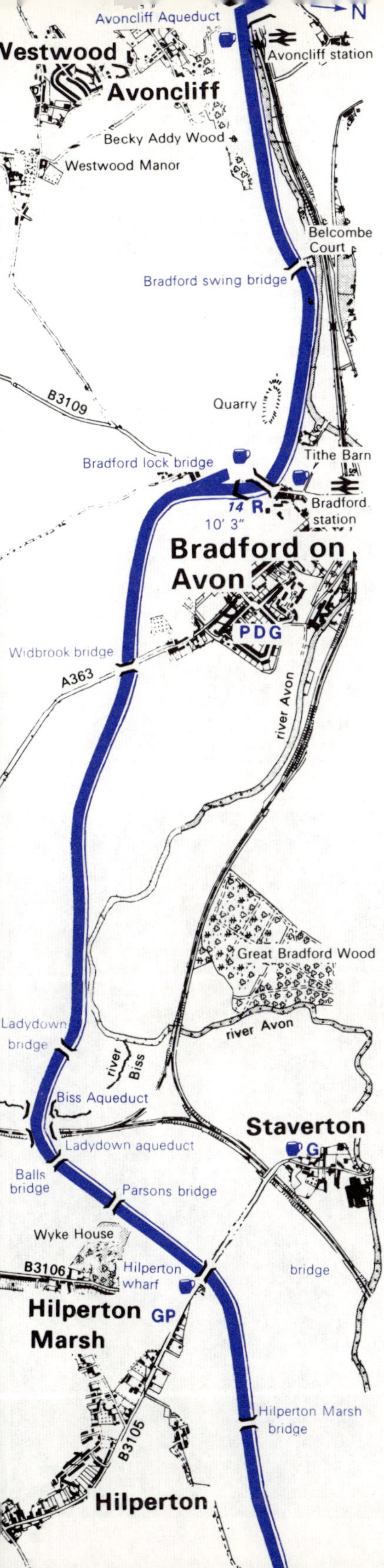

Bradford on Avon

5 miles

Continuing its westerly course, the canal passes through open pastureland: the wide Avon valley which the canal now follows begins gradually to narrow, as the hills encroach to the north and the south. The canal curves round below Hilperton; although the main village is a mile to the south, there is a convenient pub, post office and stores by the road bridge. The old wharf here is disused, although some original buildings survive. West of Hilperton the canal passes the grounds of Wyke House, whose Jacobean-style towers stand among the trees; then the land to the north falls away as the canal returns to its original course on a sweeping curve. Canal and river now converge as the canal swings on a huge embankment towards the Avon, crossing the railway and the river Biss on two stone aqueducts. The classical arch over the river is particularly handsome; it is necessary to walk down the side of the embankment in order to see it properly. The view northwards across the Avon valley is very fine. For a while river and canal run side by side, the river down in the valley, the canal high above in a side cutting, shielded by trees, and then they part again to make their separate entries into Bradford. The canal stays high above the town, which fills the steep-sided valley, while the river cuts the town in two. Bradford basin appears suddenly, followed by the lock, and then the canal turns to pass to the south of the town. It rejoins the course of the river, and now the two run closely together all the way to Bath. West of Bradford the canal passes through beautiful woods on the steep southern slope of the valley. From this point there are fine views of the town, spread out beyond the Tithe Barn, which is right beside the canal. The Avon rushes along the valley, beyond it the railway appears and disappears among the trees on the far side, while high above, the canal pursues its more sedate course towards Bath. The thick woods often give the canal user a feeling of total seclusion. A swing bridge in the middle of the wood marks the end of the section in water. At once the canal bed is dry, and overgrown; in places landslides have totally buried the canal. This dry section, which seems to belong to another age altogether, continues through the woods, and then turns northwards on to the Avoncliff aqueduct. High above the fast-flowing river Avon the canal is carried across the valley and the railway. It then turns west again to continue its wooded course. The towpath crosses back to the south side by the aqueduct.

Avoncliff
Wilts. Station. A hamlet clustered in the woods beside the canal. Originally it was a centre of weaving, and many traces of the old industry can be seen: weavers' cottages, and the old mills on the Avon, which falls noisily over a weir at this point. The hamlet is dominated by Rennie's aqueduct, built in 1804 to take the canal across the valley to the north side. A classical stone structure, the aqueduct suffered from casual repair work and patching in brick when owned by the GWR; which makes it look sad and neglected.

Bradford on Avon
Wilts. Pop 7,300. EC Wed. PO, tel, stores, garage, bank, station. Set in the steeply wooded Avon valley, Bradford is one of the beauty spots of Wiltshire, and one of the highlights of the canal. Rather like a miniature Bath, the town is composed of fine stone terraces rising sharply away from the river, which cuts through the centre of the town. Until the 19thC it was a prosperous centre for weaving, but a depression killed the industry and drove most of the workers away. More recently the town has become famous again as the home of the Moulton bicycle. Bradford is rich in architectural treasures from the Saxon period to the 19thC, while the abundance of fine 18thC

houses make an exploration of the town a positive pleasure. The centre is very compact, and so the walk down the hill from the canal wharf lays most of it open to inspection, including the town bridge, Holy Trinity Church, the Victorian town hall, and the fine gothic revival factory that dominates the riverside. There is also a swimming pool near the canal.

Bradford Wharf The canal wharf is particularly attractive. There is a small dock with some of the original buildings still standing, plenty of mooring space, and an old canal pub beside the lock.

Town Bridge The nine-arched bridge is unusual in having a chapel in the middle, one of the few still surviving in Britain. Parts of the bridge, including the chapel, are medieval, but much dates from a 17thC rebuilding. During the 17thC and 18thC the chapel fell out of use, and was turned into a small prison, serving as the town lock-up.

Holy Trinity Church Basically a 12thC building with additions dating over the next three centuries. Inside are some medieval wall paintings, and fine 18thC monuments.

Saxon Church of St Lawrence Founded in 705, this tiny church was enlarged in the 10thC. Since then it has survived essentially unchanged, having been at various times a school, a cottage and a slaughterhouse. The true origins and purpose of the building were only rediscovered in the 19thC, and so it remains one of the best preserved Saxon churches in England.

Great Tithe Barn Standing below the canal embankment, this great stone building is one of the finest tithe barns in England. It was built in the 14thC by the Abbess of Shaftesbury. Its great length (168ft) is broken by two porches, with massive doors which open to reveal the beamed roof. Maintained by the Department of the Environment, the barn now contains a collection of old agricultural implements, and is *open to the public at all reasonable times.*

Westwood Manor
1 mile SW of Bradford. This 15thC stone manor house contains much original Jacobean plaster and woodwork, although much was lost when the manor became a farm in the 18thC. Skilful restoration by the National Trust has returned the manor to its former glory. *Open: Apr-Sep. Wed 14.00–18.00.*

Staverton
Wilts. Tel, stores. The village lies north of the canal, spreading down to the banks of the Avon, where there is a small Nestlés factory. A small isolated part of the Avon is navigable here, and is used by a few pleasure boats. In the village are terraces of weavers' cottages, a sign of what was once the staple trade of the area.

Hilperton
Wilts. PO, tel, stores, garage. A scattered village that stretches away from the settlement by the canal wharf. Wyke House stands to the west of the village; this very ornate Jacobean mansion was in fact built in 1865, a replica of the original house. The house is not open to the public.

PUBS

Crossed Guns Avoncliff. 17thC gabled inn by south side of aqueduct.

Barge Inn Bradford. Canalside, by wharf. Food, B&B. Tel 3403.

Canal Tavern Bradford. Canalside.

Old Bear Inn Staverton. ¼ mile NW of canal, on B3105.

Kings Arms Hilperton. 100yds S of Hilperton Wharf.

BOAT TRIPS

'Kenavon Queen' carries 12 passengers on trips on the pound above Bradford lock. Operated by the Kennet & Avon Canal Trust. For bookings apply: 151 Trowbridge Road, Bradford, Wilts. (2129).

FISHING

Big pike, roach, bream, tench and the occasional perch can be caught in the wooded stretch to the east of Bradford.

Avoncliff Aqueduct, where the Kennet & Avon Canal crosses over the Bristol Avon.

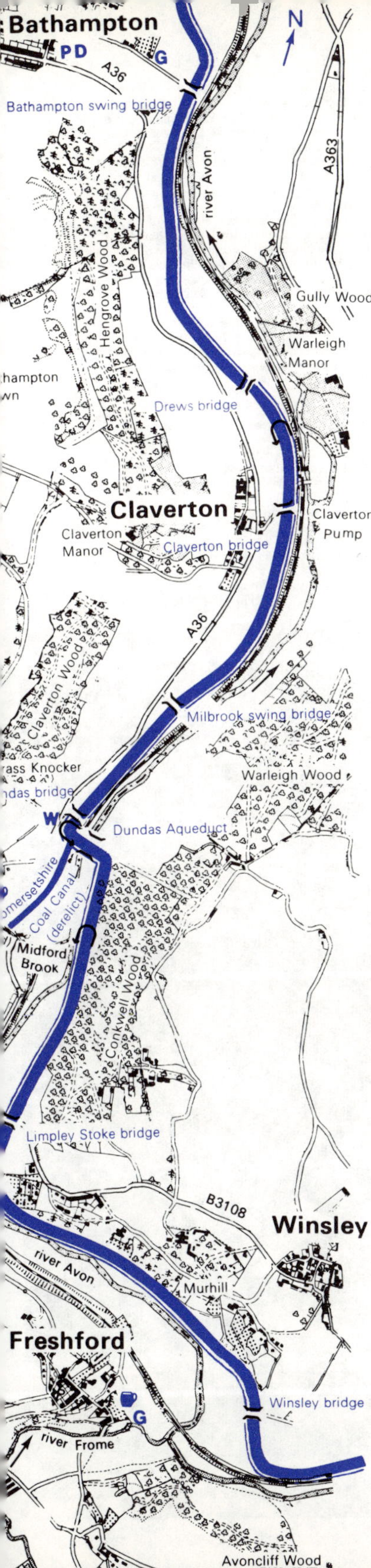

Claverton

$5\frac{1}{4}$ miles

Leaving the Avoncliff aqueduct, the dry section of the canal continues westwards through the woods above the river Avon. The valley gets steeper and narrower as it approaches Bath, and thick woods cover both sides. River and canal run side by side, the dry, overgrown canal bed contrasting with the fast-flowing river below. The canal passes Limpley Stoke, scattered over the southern valley side, and then leaves the dry section behind. The canal is in water from here all the way to Bath. The country opens out slightly, to allow views across the valley as the canal approaches the Dundas aqueduct, perhaps the most well known feature of the Kennet & Avon. Emerging from the woods, the canal turns suddenly on to the aqueduct which carries it across the Avon valley and the railway to the south side. At the southern end of the aqueduct is a small wharf and basin, with an old crane standing over the water. Here was the junction with the Somersetshire Coal Canal, which, until its closure in 1904, ran south from the Kennet & Avon Canal towards Paulton. Beyond the basin the towpath turns over to the north bank, where it remains until Bath is reached. The canal enters another thickly wooded stretch, a side cutting taking it towards Claverton. The woods soon give way to allow fine views to the north, across rolling country and the railway and river Avon in the valley below. Claverton flanks the canal, but it is hidden by the folds of the land to the south. Access is easy, and both the village and Claverton Manor are worth a visit. Claverton Ram, a water-powered pump which lifts water up from the Avon to feed the canal, is being restored by the Kennet & Avon Canal Trust, with help from engineering students from Bath University. The more open country continues, allowing views across the valley to Warleigh Manor, now a college, and to Bathford church. The canal follows the contours of the land as it turns towards Bath, maintaining the level of the 9-mile pound that runs from Bradford to Bath top lock.

Claverton
Somerset. Tel. Although devoid of all facilities, Claverton is well worth a visit. It is a manorial village of stone houses, surrounding the 17thC farm, and in earlier days clearly dependent upon Claverton Manor. The main road misses the village, increasing the peace and seclusion.

Claverton Manor
Somerset. The American Museum in Britain. The manor was built in 1820 by Sir Jeffry Wyatville in the Greek revival style. It now houses a museum of American decorative arts from the late 17thC to the mid-19thC. *Open: Apr to Oct Tue to Sun, 14.00–17.00. Winter on application only.* (Bath 60503).

Claverton Pump The waterwheel pump at Claverton is the only one of its kind on British canals. Designed by John Rennie, the pump was built to feed the nine-mile Bradford-Bath pound, and started operating in 1813. The two undershot breast wheels, each 15ft in diameter and 11ft wide, then powered the pumping machinery until a major breakdown in 1952 prompted its closure, and replacement by a temporary diesel pump. In 1969 a programme of complete renovation of the pump and its machinery got under way, and soon it will be raising water once more 53ft from the Avon to the canal. With Crofton and Claverton, the canal will possess two valuable working industrial monuments.

Dundas Aqueduct
Built in 1804, this three-arch classical stone aqueduct is justifiably one of the most well-known features of the canal, and stands as a fitting monument to the architectural and engineering skill of John Rennie. It is necessary to leave the canal and walk down into the valley below to appreciate the beauty of the aqueduct, and see it in the

context of the narrow Avon valley into which it fits so well.

Somersetshire Coal Canal
Opened in 1805, this narrow canal was sponsored by the Somerset Coal owners, who wanted a more efficient means of moving their coal to Bath, Bristol and to the rest of England. Originally surveyed by Rennie in 1793, the canal was to run from Limpley Stoke to Paulton, with a branch to Radstock. There were steep gradients to overcome at Midford and Combe Hay, and these plagued the canal throughout its life. The Radstock arm was never completed and tramroads were built over the difficult stretches. The canal was never profitable, and was sold to the Somerset & Dorset Railway in 1871. The main line was completed throughout, but not before some remarkable solutions to the problems of the Combe Hay gradient had been tried out. First there was Robert Weldon's caisson lock; a watertight caisson, large enough to hold a narrow boat and crew, was pulled up and down a water-filled cistern 88ft deep by means of a rack and pinion. This terrifying device was soon replaced by an inclined plane, which in turn was replaced by a conventional flight of locks. Once open, the canal carried a large tonnage of coal throughout the 19thC; it served 30 collieries more directly than the railway. However by the end of the century the inevitable competition was taking away the traffic, which finally stopped in 1898. The canal was officially abandoned in 1904. The main line can still be traced on foot by careful explorers, but the arm has disappeared. *(See 'The Somersetshire Coal Canal and Railways' by K. R. Clew, published by David & Charles.)*

Limpley Stoke
Wilts. PO, tel, stores. Built on the side of the valley overlooking the river. Limpley Stoke is a quiet village, a residential outpost of Bath. The little church includes work of all periods, from Norman to the 20thC; inside is a collection of carved coffin lids..

Freshford
Somerset. PO, tel, stores, garage. Although not on the canal, Freshford is well worth the ½-mile walk south from Limpley Stoke. It is a particularly attractive village, set on the side of the steep hill that flanks the confluence of the rivers Avon and Frome. At the top of the hill is the church, and terraces of handsome stone houses fall away in both directions, filling the valley below, and crowding the narrow streets. At the bottom of the hill is the river, crossed by the medieval bridge.

PUBS

Viaduct Hotel Claverton, on Bath road 200yds S of Dundas Aqueduct. B&B. Restaurant. (Limpley Stoke 3187).

Old Boathouse Restaurant Limpley Stoke. Beside Avon bridge. (Limpley Stoke 3150).

Hop Pole Limpley Stoke.

FISHING

Limpley Stoke and Claverton are good fishing lengths, and specimen pike, roach, bream and tench can be found. A 2-lb roach, and an 18¼-lb pike have been caught here.

Widcombe locks, Bath. The lock here has been completely restored: work continues on repairs to the waterway wall.

Bath

5¾ miles

Following the course of the river Avon, the canal turns west towards Bath, leaving behind Bathford church on the opposite side of the valley. Groups of houses appear more frequently scattered among the trees of the Avon valley; these form the outposts of Bath whose suburbs are now visible to the west. The canal passes through Bathampton, on a low embankment above the school and church, and then continues on a straight course, closely flanked by the railway, which is in a cutting below. On the south bank there are gardens running down to the water, which accompany the canal into Bath. The entry into Bath is magnificent. The canal sweeps round the south of the city, cut into the side of the hill, and so there are extensive views across Bath. From this point it is possible to pick out many of the features of the city, and the Georgian terraces can be seen spread out over the far side of the valley. As the buildings fill the valley, canal and river Avon part, to make their separate entries into Bath. The first Georgian buildings flank the canal as it reaches Sydney Gardens. A short tunnel with a fine Adamesque portal takes the canal under a road, and then it passes two pretty cast iron bridges, both dated 1800. A cutting carries the canal through this attractive part of Bath, and so the houses seem to hang over the water. Another ornamental tunnel actually carries houses over the canal, among them Cleveland House, the old canal company's headquarters. The towpath turns over briefly to the south side, returning to the north at the next bridge. The cutting then ends, once more allowing magnificent views over the city before the canal reaches lock 13, the top lock of the Widcombe flight. This flight of 7 locks which takes the canal down to join the Avon is being restored at present. Locks 8 and 9 are to be merged together as part of a road building scheme, making one new lock with a fall of over 19 feet. It is hoped that the whole flight will be reopened by the end of 1973, thus making possible navigation from Bristol to the Dundas Aqueduct. At present the towpath turns over to the south bank by lock 9. The canal joins the Avon immediately beyond Bath lower lock (number 7) in the middle of the industrial quarter of Bath. The railway station is opposite the junction of canal and river, and factories and warehouses flank the Avon as it leaves Bath. The fine Georgian city surrounds the unnavigable Avon to the east. The junction is the best point of access for Bath as a whole. New roads have made the towpath of the Avon Navigation difficult to find, but after a while it establishes itself on the north bank. A long belt of industry accompanies the river out of Bath but access to the towpath is always easy. There are several footbridges across the river, some of them private, and a disused railway crosses twice as the river meanders in long, gentle curves. As the industry gradually falls away, the river divides; the right fork leads to Weston lock, the left to a weir, as the river continues its fall to the sea.

Navigating the Bristol Avon
Pleasure boats should always give way to barges, and should let them use the locks first. In general, downstream traffic has right of way, especially through bridges. All the locks are accompanied by weirs, and so boatmen should take great care to turn into the lock cuts, and avoid the weir channel. Remember that a river always has a current, and is liable to changes in speed and level of flow. When mooring, allow enough slack on lines. Do not moor in lock cuts or near weirs. All pleasure boats should moor up at night, and show a white light. With the exception of Hanham, the locks are not manned. Remember that boats should always be held by ropes while the locks are being operated, for there is a strong flow in these large locks.

Bath
Somerset. Pop 85,000. EC Thur, MD Wed. Bath was first developed by the Romans as a spa town and resort because of its natural warm springs. They started a trend of bathing and 'taking the waters' which survives today. There are extensive Roman remains to be seen in the city, not least the baths themselves. The city grew further during the medieval period, when it was a centre of the wool trade; the fine abbey dates from this time. But the true splendour of Bath is the 18thC development, when the city grew as a resort and watering place that was frequented by all levels of English society, from Royalty downwards. Despite heavy bombing in the 1939–45 war Bath is still a magnificent memorial to the 18thC, and neo-classicism generally. The terraces that adorn the steep northern slope of the Avon valley contain some of the best Georgian architecture in Britain. Much of the city was designed by John Wood the Younger, who was responsible for the great sweeping Royal Crescent. Other architects include Thomas Baldwin, who built the Guildhall, 1766–75, and the Pump Room, 1789–99, and Robert Adam, whose Pulteney Bridge carries terraces of shops across the Avon. Plagued by traffic, Bath is best seen on foot, for its glories and riches are far too numerous to list. Visitors should not fail to try the waters, which gush continuously from a fountain outside the Pump Room.

Bath Abbey
Set in an attractive piazza, the abbey is a pleasingly uniform perpendicular building, founded in 1499. Twin towers crown the west front, decorated with carved angels ascending and descending ladders. Inside, the abbey is justly famous for its fan vaulting which covers the whole roof of the building, but which is not all of the same date. Inside also is a wealth of memorials of all periods, an interesting indication of the vast range of people who, over the ages, have come to die at Bath.

Holburne of Menstrie Museum
Great Pulteney Street (3669). Housed in an 18thC Palladian building that was designed as part of the Sydney pleasure gardens, it contains collections of silver, ceramics and 18thC paintings and furniture. *Open: weekdays 11.00–13.00 and 14.00.–17.00, Sun 14.30–17.30.*

Museum of Costume
Assembly Rooms. (28411). Display of fashion from the 17thC to the present day; one of the largest collections of costume in the world. *Open: winter weekdays 10.00–17.00, Sun 14.00–17.00. Summer weekdays 9.30–18.00, Sun 11.00–18.00.*

Bath Roman Museum
Abbey Churchyard. (28808). The great bath buildings with their dependent temple were the centre of Roman Bath. Much of these survive, incorporated into the 18thC Pump Room. The museum, attached to the bath buildings, contains finds excavated from the site. *Open: daily 9.00–18.00 (winter 17.00).*

1 Royal Crescent
A typical mid-18thC house, complete with original furniture and fittings. *Open: Mar-Oct weekdays 11.00–17.00, Sun 14.00–17.00.*

Victoria Art Gallery
Bridge Street. (28144). Collection of 18thC and modern paintings, prints and ceramics. Visiting exhibitions. *Open: weekdays 10.00–18.00.*

Bathampton
Somerset. PO, tel, stores, garage. The centre of the village surrounds the canal, and is still compact and undeveloped; but new housing around the village has turned it into a suburb of Bath. The church, well placed in an attractive churchyard, is mostly 19thC.

BOAT TRIPS

'Jane Austen' carries passengers between Bathampton and Bath, offering scenic views of the city. For bookings contact Commander C. Wray-Bliss, Dundas Carrying Co., Box Hill, Wilts. (Box 665).

PUBS

Plenty of pubs in Bath.

Trattoria da Pietro 39 Gay st, Bath. (27919). A lively restaurant in the centre of town serving Italian food. Taped music and singing waiters. *Lunch 12.00–14.15, dinner 18.00–23.15. In the Good Food Guide since 1967.*

Hole in the Wall 16 George st, Bath. (25242). Original dishes are cooked to a standard sufficiently high to have been awarded a laurel wreath by the Good Food Guide. Prices are reasonable considering the quality of such specialities as Mexican ceviche, Dublin Bay prawns in cream and brandy, duckling farci à la mode ancienne de Provence and St Emilion au chocolat. *Lunch 12.15–14.15, dinner 18.30–22.00 (Closed Xmas to New Year). Must book. In the Good Food Guide since 1953.*

Bruno's 2 George st, Bath. (25141). Across the road from the 'Hole in the Wall', this friendly restaurant serves home-made soups and pasta. Other dishes include a roast suckling pig with sauerkraut. *Lunch 12.15–14.15, dinner 18.30–22.00. (Closed Mon, Sun.) In the Good Food Guide since 1969.*

George Inn Bathampton. Canalside. Food.

CARAVANS & CAMPING

Newbridge Caravan Park Bath. (28778). Just S of New Bridge. Overnight stops for touring caravans and motor caravans. Gas, showers (h&c). *Open: Mar-Oct.*

YOUTH HOSTELS

Bathwick Hill, Bath. (22626).

FISHING

As the waterway is not continuous at present, the fishing is confined to certain areas. Several of these are good, the quality of the fishing being improved in several places by the isolation created by disused locks. At Bath the canal is fairly well stocked with coarse fish, and many popular species can be caught.

TOURIST INFORMATION CENTRE

Bath Abbey Churchyard, Bath. (28411, or 28891 on Sats).

Widcombe locks, Bath.

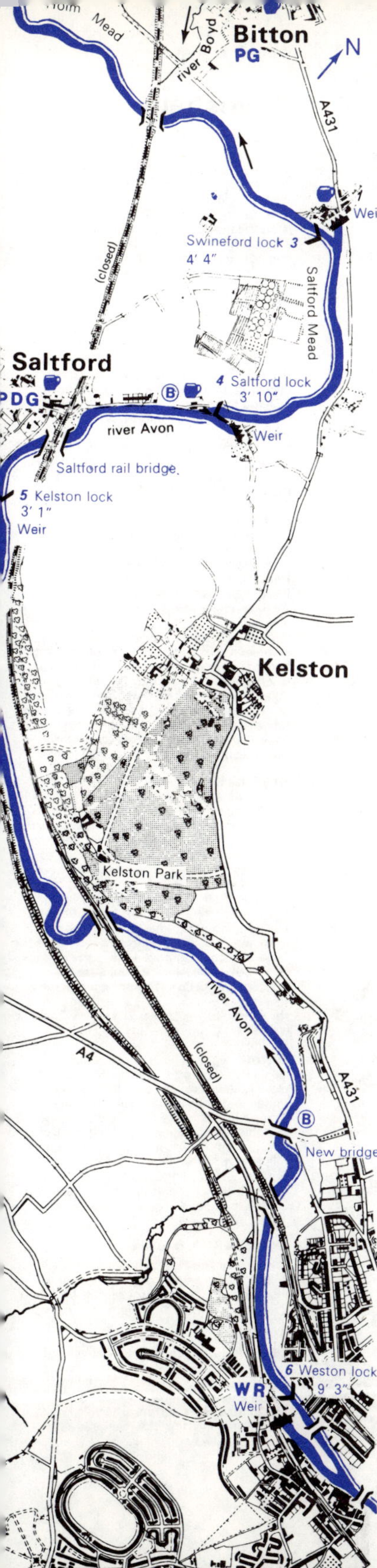

Saltford

6 miles

The river Avon at last leaves behind the industries of Bath, and enters a wooded stretch. The railway closely follows the south bank, vanishing at one point into a tunnel. As the river continues its wide, wandering course westwards, the valley opens out, and rolling hills and pastureland flank both banks. A disused rail bridge is followed by the elegant single stone arch of New Bridge, carrying the A4. At this point the towpath crosses to the south bank, although, as on all river navigations, its position is never well-defined. After New Bridge there is a small boatyard, and a line of moored craft along the north bank. The river passes Kelston Park in a series of gentle bends, against a background of wooded hills to the north. The disused railway crosses the Avon for the fourth time, an indication of the river's winding course. It is advisable for walkers to cross the river by this rail bridge, for the better path is on the north bank. The river straightens as it approaches Kelston lock where the stream again divides. Boatmen should take the right fork to the lock, and avoid the weir on the left. Saltford can be seen among the trees on the south bank. Mooring is possible by the lock, and this is the best access point for boatmen visiting the town. Walkers should continue along the bank, and again cross the river by the disused rail bridge. After the bridge a path leads down to the village. The towpath continues along the north bank, as the river curves towards Saltford lock. At one time there was a horse and passenger ferry near the lock which allowed the towpath to cross to the south bank. The ferry is no more, and so it is very difficult nowadays to make an uninterrupted riverside walk. By Saltford lock is Saltford Sailing Club where ice creams and sweets can be bought; there is also a riverside pub. At Saltford the lock is on the left. The river then turns past Saltford Mead towards Swineford, passing a large factory on the low-lying land to the south. At Swineford the river again divides, the left fork leading to the particularly attractive lock, which is set against a background of trees and old mill buildings. The better towpath remains on the south bank, though this means that walkers cannot get to Swineford village.

Swineford
Glos. PO, tel, stores. Although bisected by the A431, the settlement by the river is still attractive; old mill buildings overlook the long weir.

Saltford
Somerset. Pop 1,700. EC Wed. PO, tel, stores, garage. Although Saltford has been developed as a large-scale dormitory suburb, the older parts by the river are still pretty and secluded.

Saltford Manor
Situated by the church, the Manor is one of the oldest inhabited houses in England. Much of the building is still original Norman work, but it is hidden behind a 17thC façade.

BOATYARDS & BWB

Ⓑ **Bristol Boats** by Saltford lock. Jolly Sailor Boatyard, Mead lane, Saltford, Somerset (2032 and 3401). Boat building, sales and repairs, inboard and outboard engine sales and repairs. Slipway available. Chandlery.

Ⓑ **Newbridge Boating Station** Newbridge road, Bath. (24301). Water, chandlery, dustbins. Slipway (up to 12ft), moorings, winter storage. Boat building, sales and repairs, inboard and outboard engines repaired. Skiffs and motorboats for hire by the hour.

PUBS

Swan Swineford. Access is easy for boats, but virtually impossible for towpath walkers.

Jolly Sailor By Saltford lock. Riverside garden.

Bird in Hand Saltford.

Kennet & Avon

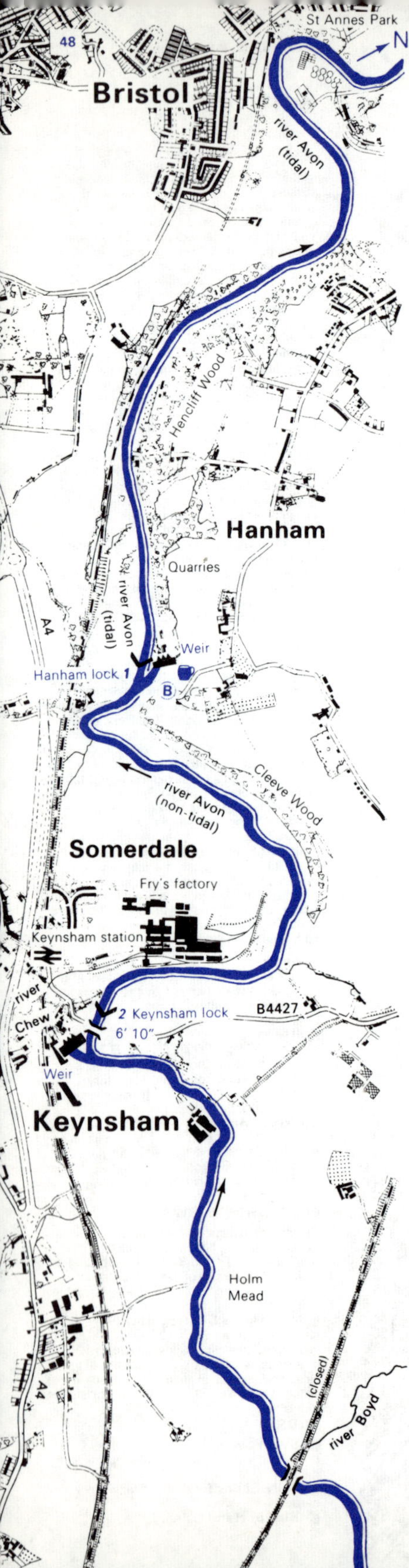

Keynsham

6½ miles

As the river passes Bitton it reaches another disused railway bridge. Once again towpath walkers should use this bridge to cross the river, for from here to Bristol the better path is on the north bank. Low-lying pasture and arable land continue to flank the river in its meandering course towards Keynsham. After passing a vast brick and stone factory complex, dated 1881, which dominates the south bank, the river starts a long horseshoe bend that leads to Keynsham. The river divides, the right fork leading to the lock. Keynsham lies well to the south of the river, but is easily accessible. There is a small settlement round the lock, rather overawed by the industry that surrounds it. Leaving Keynsham, on the south bank there is a huge chocolate and sweet factory, which fills the air with a heavy, sweet smell. Incongruously buried beneath the factory is the site of a Roman villa. As the valley narrows, steep wooded hills return to follow the north bank of the river as it twists and turns. After a particularly sharp bend, the southern hills approach as well, and Hanham lock appears. Again the river divides, the left fork leading to the lock. This is lock 1, the last lock between Reading and Bristol, and the end of the BWB's jurisdiction. Note that the river Avon is tidal west of Hanham lock. There is a small hamlet on the north bank, overlooking the weir, and a pub and boatyard. In summer a passenger ferry operates. From here the river Avon continues through a steeply wooded valley to Bristol; a canal takes boats through Bristol harbour and then the navigation rejoins the river which flows down to join the Severn estuary at Avonmouth, having passed the Clifton Suspension Bridge and the Avon gorge.

Keynsham
Somerset. All services. Keynsham has grown steadily along the Bristol road, and so is now a vast shapeless suburb. However, the centre still retains a feeling of independence, and has many traces of Keynsham's past. An Augustinian Abbey was founded here in 1170, and there are a few surviving remains in Abbey Park. Elsewhere in the main street are a few 17thC and 18thC houses, but bungalows and modern shops predominate. The main feature of interest is the large church. Originally 13thC, the interior is now attractively Victorian, after the restoration of 1861. The ornamental west tower was built in 1734, after the earlier tower was blown down in a storm. There is a fine 16thC monument to Sir Henry Bridges.

Bitton
Glos. PO, tel, stores, garage. Although a main road village, Bitton's heart survives intact south of the road. Here is a fine group formed by the church, the grange, and the 18thC vicarage, all built around the churchyard. The church is very splendid; it has a long Saxon nave with Norman details, a 14thC chancel, and a magnificently decorative late 14thC tower. On the main road is an attractive early 19thC Wesleyan chapel.

BOATYARDS & BWB

Chequers Inn Hanham. (Bristol 674242). Water, dustbins and slipway available, overnight moorings. Small boats for hire by the hour or day.

PUBS

Chequers Inn Hanham. Riverside, overlooking the lock.
X **Avon Grill** Keynsham, by the lock. Food. Plenty of pubs in Keynsham.
White Hart Bitton. On A431, most easily reached from the river along the disused railway.

Monmouthshire & Brecon

Maximum dimensions

Length: 55′
Beam: 8′ 6″
Headroom: 5′ 7″

Mileage

PONTYPOOL to
Goytre Wharf: 6
Llanfoist: 11½
Gilwern: 14½
Llangattock Bridge: 18
Talybont: 26½
BRECON: 33¼

Total 6 locks

In 1792 the Act of Authorisation for the Monmouthshire Canal was passed. This gave permission for a canal to be cut from the estuary of the river Usk at Newport to Pontnewynydd, north of Pontypool. In addition to this 11-mile main line, there was to be an 11-mile branch from Malpas to Crumlin. The canal was designed to connect with a large network of tramways that were to be built to serve the iron ore, limestone and coal mines of the area. Thomas Dadford was appointed engineer, and the canal was opened in 1796, with many of the tramways still to be built.

The close relationship between canal and tramway from the start of the scheme was a feature of South Wales. The promoters of the canal saw these embryo railways as a means of increasing their revenue, without suspecting that they were encouraging the development of a means of transport that later was to cause the downfall of the canals.

When the Act for the Brecknock & Abergavenny Canal was passed in 1793, it was conceived in very similar terms to its southerly neighbour. The Canal was planned to connect Brecon with the river Usk at Caerleon, to serve as a link between the various tramways and the Usk navigation. The directors of the Monmouthshire Canal persuaded the promoters of this rival venture to alter their plans to include a junction with their own canal, whose construction was well under way by this date. And so the Brecknock & Abergavenny Canal, with Thomas Dadford again as engineer, was cut from Brecon to Pontymoile basin, where it joined the Monmouthshire Canal. Construction was begun in 1797. It progressed slowly because the company's first priority was the building of the tramways, a more immediate source of revenue. The canal was only to be built when all the tramways were in operation, to serve as a keystone for the whole system. After the usual delays caused by shortage of money, the Brecknock & Abergavenny Canal was opened throughout in 1812.

For a while the two canals were profitable, because the iron and coal cargoes justified the use of both canal and tramway. However, the greater speed and efficiency of the railways was soon apparent, and by the 1850s there were many schemes to give up the canals, and rely entirely on the rail system. Some were put forward by the canal companies, in order to protect their interests as best they could. In 1865 the Monmouthshire and the Brecknock & Abergavenny Canal Companies amalgamated, becoming the Monmouthshire & Brecon Canal Company; but already it was too late for the merger to be effective. Revenues were dropping fast as the railway tentacles reached through the South Wales coalfields. Later the whole system was bought by the Great Western Railway, and by the turn of the century only a few boats were still using the canal. Bit by bit the original Monmouthshire Canal was closed, but the Brecon line was kept open as a water channel. In 1962 the network was formally 'abandoned', and parts were filled in.

However, with the development of the Brecon Beacons National Park, the amenity potential of the Brecon line was realised. In 1964 the slow task of restoration was begun by BWB, with the help of Brecon and Monmouth County Councils.

The locks were restored, and soon boats were once more able to cruise from Pontymoile to Talybont. In 1970 the low fixed bridge at Talybont was replaced with a new lifting bridge, and once again navigation was open all along the old Brecknock & Abergavenny Canal. The Monmouthshire line from Pontymoile to the Usk at Newport can never be reopened in its entirety, for stretches have vanished completely; but there is a possibility that it may be restored as far south as Cwmbran, where it would prove a valuable asset to the developing new town.

Natural history

A diversity of interesting and colourful wild plants and animals can be seen along the Monmouthshire & Brecon Canal. For much of its length the canal is tree-lined, mostly by alders which can be recognized by their smooth, roundish leaves with jagged edges and their clusters of little cones. Interspersed with the alders, or in small copses nearby, is a variety of other trees such as oak, ash, elm and sycamore. Between Llanfoist and Govilon fine beech trees clothe the hillside and reach down to the canal, and in spring wild cherries are conspicuously beautiful in blossom between Llangynidr and Crickhowell.

The smaller aquatic plants grow best in less shady places. Rooted in the mud at the bottom of the canal, and completely submerged in water, are the true aquatics like the feathery-leaved water milfoil and Canadian pondweed. The latter, a North American plant, spread rapidly after its introduction into this country halfway through the last century. Other water plants to be seen include bur-reeds, with long ribbon shaped leaves floating on the surface, and water plantains with oval leaves thrust above the water and tiny pink or white three-petalled flowers.

Bordering the canal grow many gaily coloured marsh plants, in bloom from July until Autumn. Especially conspicuous are the tall 'codlins and cream', with rose flowers, and hemp agrimony with fluffy heads of pink flowers. The sweetly scented, feathery clusters of cream meadowsweet contrast effectively with the lovely blue of water forget-me-not. Marsh woundwort with its lilac flowers is frequent and here and there along the banks the blue trumpets of skullcap can be seen.

The animal life is also rich and varied. The kingfisher with its scintillating blue-green plumage is always an exciting sight. Fortunately it is now fairly frequent again along the canal. Another bird which comes to fish is the tall grey heron which flaps away on slow wingbeats when disturbed. The largest bird likely to be encountered is the mute swan; a family party including four or five cygnets may often be seen near Brecon. Moorhens can be observed at several spots, either swimming along in their inimitable jerky style or walking about on the banks in search of food. Where the canal overlooks the River Usk as at Llanhamlach or crosses it via Brynich Aqueduct, the dipper—a dark, tubby, thrush-sized bird with a white front—may be seen bobbing up and down on stones in the river. Although its main habitat is the fast-flowing river, it may forsake this for the canal on occasions and be seen on the towpath. In summer, from bramble thickets and hedgerows along the towpath issue the songs of whitethroats and garden warblers. Many small birds such as tree creepers, nuthatches and tits feed in the overhanging trees, and in winter the alders are sometimes thronged with twittering parties of siskins, searching the cones for seeds.

Of the wealth of smaller animal life along the canal, the dragonflies are perhaps the most noticeable, as they patrol to and fro over their particular stretch of water. Occasionally the female dragonflies may be seen dipping their long abdomens into the water at intervals to lay eggs on the submerged water plants. Pond skaters are the numerous small dark insects with long legs which dart away over the surface of the water at your approach. On marsh plants like the fragrant water mint, beautiful beetles with a greenish metallic lustre may be found. During May and June, orange-tip butterflies are on the wing along the banks, where the females (with grey, not orange-tipped wings) lay eggs on milkmaid plants on which their caterpillars feed. Later in summer, speckled wood butterflies may also be seen flying or basking in sunny glades.

Brecon Beacons National Park

The Park covers 519 square miles of mountain and hill country, embracing parts of the Herefordshire border, north Monmouthshire, south Breconshire and east Carmarthenshire. Apart from a great variety of fine scenery, the Park also includes three nature reserves, a forest reserve, opportunities for fishing, caving, pony trekking, sailing and boating, and several towns of interest to tourists, notably Brecon, Crickhowell, Talgarth and Hay-on-Wye; in addition Abergavenny, Llandovery and Llandeilo are just outside the Park boundary. All the canal is within the Park, the only one in Britain contained wholly in a National Park—a factor that greatly strengthened the case for its restoration and re-opening. The canal is an excellent introduction to the Park, crossing it roughly from south-east to north-west; in several places there are foot and bridle paths leading away into the mountains from the towpath. Various main roads cross the Park, and so access by car is easy, but the whole area is best explored on foot. A good place to start any exploration is the Mountain Centre, 1,000′ up on Mynydd Illtud, above the village of Libanus, 4 miles south west of Brecon. There are rest and refreshment rooms, car parks and picnic sites overlooking the Brecon Beacons, and the Wardens at the Centre give lectures to visitors and youth groups on the Park and the use of the countryside.

Tourist Information Centres in the National Park

Monk street, Abergavenny (3254).
6 Glamorgan street, Brecon (2763).

Mountain Centre

Libanus, Brecon (Brecon 3366).

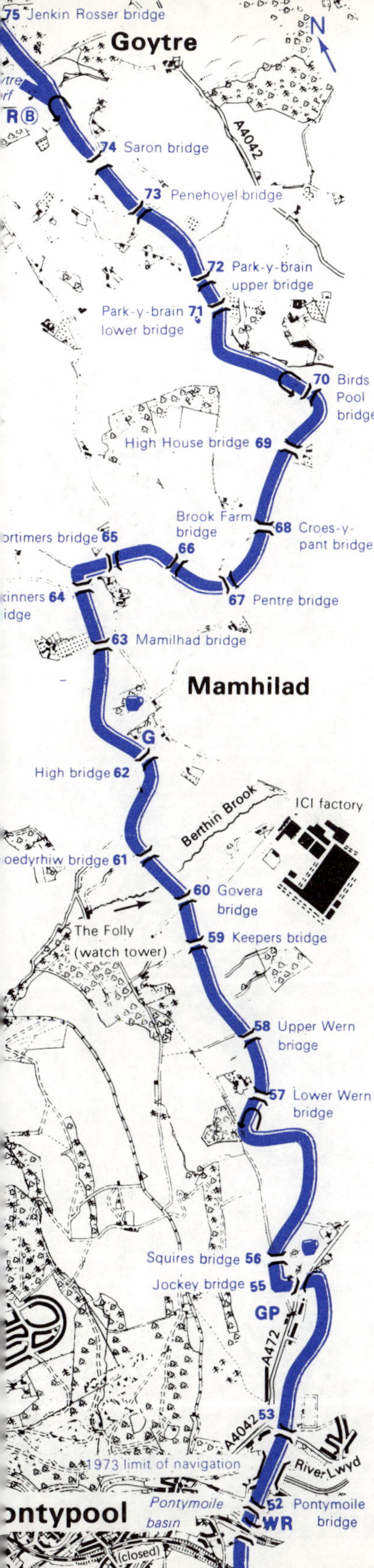

Pontypool

6 miles

Before the closure of the Newport section, the old Monmouthshire & Brecon Canal used to run from Brecon to the estuary of the river Usk at Newport, and thus to the sea. After a quarter of a century of decay, the upper section, from Pontypool to Brecon, was restored and reopened to craft in 1970. Although Jockey bridge, Pontymoile, is the BWB official limit of navigation, the canal continues as far south as Crown Bridge in Sebastopol. This low culvert bridge cuts off the now abandoned section to Newport, although there is a possibility of eventual restoration as far as Cwmbran locks. Passing the turn-over bridge, toll house and the site of the stop lock that marked the junction with the southern section, the canal turns north-east, on to an embankment that carries it past the Pilkington Glass works, over the river Lwyd on an aqueduct, and under the A4042 road bridge. This is the most convenient access point for Pontypool, Pontymoile and the railway station (Pontypool Road). After a few canalside gardens, all traces of the town are left behind, and the canal starts its meandering contour course towards Brecon. The character of this canal is quickly apparent; it twists and turns, clinging to the hillside on the west, while to the east wide views open up across the rolling pastures and woods of the Usk valley. Apart from the huge ICI fibres factory which dominates the valley throughout this section, the canal is entirely quiet, rural and isolated. The winding course, and the frequent stone bridges (all clearly numbered) make the canal interesting, for every bend offers a different view of the steep hills to the west and the valley to the east. Navigators should look out for the cast-iron mile posts which survive irregularly along the length of the canal. There are no villages by the canal in this section, but services and pubs are never more than a short walk away, at Pontymoile, at Mamhilad and at Penperlleni. Main roads also keep their distance, although they are generally clearly visible in the valley below the canal. After passing the long tunnel-like Saron bridge (no. 74), the seclusion is interrupted by long lines of moored boats, which are the prelude to Goytre Wharf, now the home of Red Line Boats. Old lime kilns can be seen by the wharf, indicating the agricultural nature of the canal in its heyday. The wooded hills sweep down to the wharf, giving it a most attractive setting.

Penperlleni
Mon. PO, tel, stores, garage. Main road village useful for supplies.

Mamhilad
Mon. Tel, stores. A little hillside hamlet scattered round the church. Overshadowed by massive yew trees, the pleasantly overgrown churchyard has fine views across the valley.

Pontypool
Mon. Pop 37,000. EC Thur. MD Wed, Fri, Sat. Pontypool has been an industrial town since Roman times, concentrating on the production of iron. This reached a peak in the 18thC and 19thC, but has declined in recent times. In 1720 tinplate was produced here for the first time in Britain, and in the 19thC the town was a centre for japanning—the coating of objects with an extract of oils from coal, so producing a black varnish similar to Japanese lacquer. Coal mining has also been important. Despite this industrial heritage, Pontypool has always remained a farming centre, and so the hard industrial elements are softened by the traditions of a rural market town. The steep walls of the Lwyd valley have also limited the growth of the town, and made the centre very self-contained.

Pontypool Park
Originally the seat of the Hanburys, the famous iron and steel family, this Georgian mansion is now a school. The park is open

to the public. The magnificent wrought iron entrance gates at Pontymoile (by bridge 53) were given to John Hanbury by Sarah Churchill, Duchess of Marlborough.

BOATYARDS & BWB

Ⓑ **Red Line Boats** Goytre Wharf, Pontypool, Mon. (Nantyderry 516). Hire cruisers. Water, diesel, gas, chandlery, dustbins. Repairs, outboard engine sales and repairs.

PUBS

Goytre Arms Penperlleni. (376). ¼ mile E of bridge 72. Food.

Star Mamhilad. 200yds E of bridge 62.

Horse & Jockey Pontymoile, on A472. 100yds E of bridge 55.

X **Clarence Hotel** Pontypool. (3050). Food, B&B. Plenty of pubs in Pontypool, also fish and chips.

FISHING THE MONMOUTHSHIRE & BRECON CANAL

24 miles of canal are available for fishing, from bridge 62 to bridge 126, and from bridge 134 to Brecon. Recently the canal has been extensively restocked with carp and bream. These new additions seem to have settled in well, and good weights have been reported by local anglers. A new match record was established in 1972. Day tickets or season permits are necessary, and these are obtainable from: BWB Office, Govilon Wharf, Abergavenny. (Gilwern 328). Permits for fishing from boats (valid for three weeks) may be obtained from local boat hirers.

Gentle exercise along the 'Mon & Brec'. *Derek Pratt.*

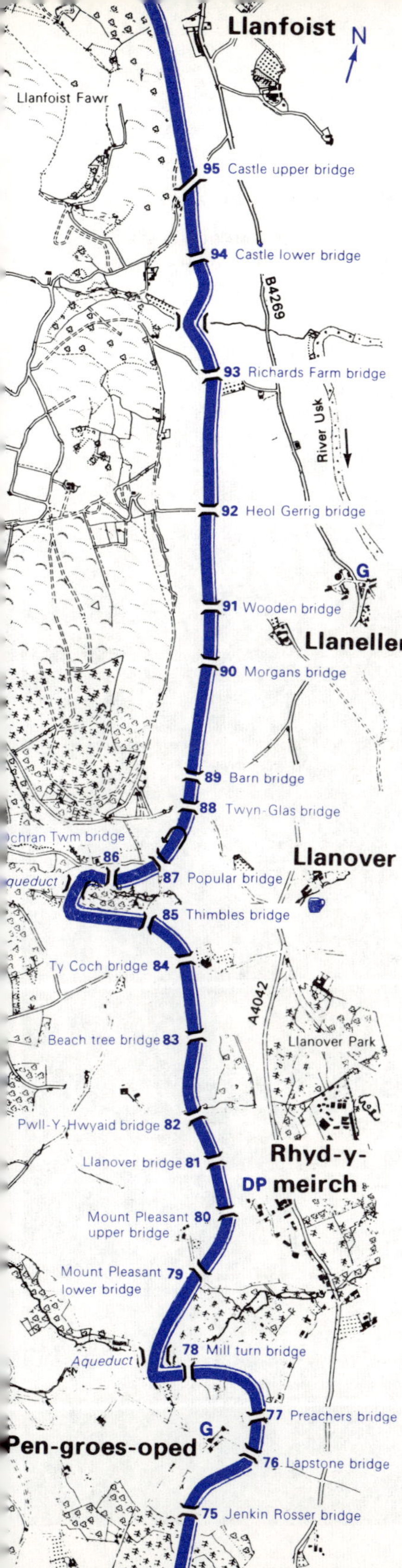

Abergavenny

$5\frac{1}{4}$ miles

Beyond Goytre Wharf, the canal continues its meandering course northwards, passing through a thick wooded cutting before returning to the pattern set in the previous section: steep wooded hills to the west, the wide valley rolling away to the east. The canal clings to its contour line high on the side of the hills, at times making horseshoe bends to avoid any change in level. At the apex of each such bend, there is generally a small stone aqueduct taking the canal over a stream that tumbles noisily down towards the valley. In several places these streams serve as feeders for the canal, half their water joining the canal, half passing beneath an aqueduct. After the first horseshoe bend, set among thick woods, the canal passes Llanover Park to the east; the house is out of sight, shielded by trees, but the village can be seen nestling in the hillside. The next bend is much wider; looking back the course of the canal along the side of the hill can be clearly seen. After the bend there is a long straight for over a mile which carries the canal through the trees above Llanellen. The river Usk is coming nearer all the time, and its course into Abergavenny is visible below. To the west the hills are now very steep, rising sharply to over 1800 feet, at times almost vertically away from the canal. From its elevated position there is an excellent view of Abergavenny, laid out like a model in the valley below, nearly a mile from the canal. In the 19thC the hills to the west were heavily mined and quarried, and many tramways constructed to carry the coal, iron ore and limestone down to the canal to be loaded into boats. There are still traces of at least ten of these tramways; often their course into the hills can be followed from the canal bridges. A good example leaves the canal at Llanfoist, by the boathouse. The loading wharves, cut into the steep hillside, can also be seen in many places along the canal.

Abergavenny
Mon. Pop 9,600. EC Thur. MD Tue. All services. Abergavenny lies beside the fast flowing river Usk, surrounded on all sides by mountains and hills; the Sugar Loaf, Blorenge and the Skirrids overlook the town, a dramatic natural wall ranging up to 2,000ft. Abergavenny enjoys this magnificent setting, living up to its reputation as the gateway to Wales. Primarily a market town, it contains the traditional mixture of buildings of all periods and styles, from the Tudor houses in the main street to the red stone of the 19thC Gothic town hall.

Abergavenny Castle
The mound of the castle dominates the town. Built in the 11thC, the castle was the scene in 1177 of a treacherous massacre of several notable Welsh leaders; invited to the castle to dine by William de Braose, they were put to death with no warning. In this violent way William made sure of his control over the surrounding lands. Parts of walls, towers and a gateway survive. Ruins and grounds *open daily*.

St Mary's Church
Originally the chapel of the Benedictine priory, the church was extensively rebuilt in the 14thC, after the destruction of the priory. It contains fine wooden 14thC choir stalls, a wooden figure of Jesse, and rich monuments in the Herbert chapel.

Abergavenny and District Museum
Situated in the castle precincts, the museum contains items of local interest, Roman coins, tools and examples of local crafts. *Open daily.*

Sugar Loaf
A conspicuous landmark 2 miles north west of Abergavenny, so named because of its flat shape. 2,130 acres, including the 1,955ft summit, are owned by the National Trust.

Rural Crafts Museum Llanvapley, 4 miles E of Abergavenny, on B4233. 500 tools, rural and farm equipment and implements, domestic bygones. *Open: Sun only all year, 15.00–18.00.*

Llanellen

Mon. PO, tel, stores. Although modern housing has greatly extended Llanellen into a suburb of Abergavenny, it is still an attractive village, especially by the 3-arch stone bridge over the Usk. The 19thC church is pleasantly set among woods.

Llanover

Mon. PO, tel, garage. The famous tower Big Ben in Westminster was named after the politician Benjamin Hall (Lord Llanover), who was responsible for its construction whilst Chief Commissioner of Works. He also initiated the tramway from Buckland House Wharf to Rhymney Ironworks, east of Talybont reservoir. The estate village is particularly elegant; stone cottages and terraces all built in the same style, pleasingly laid out with generous grass verges and trees.

PUBS

Many pubs and restaurants in Abergavenny.

Goose & Cuckoo Llanover. Off the A4042.

TOURIST INFORMATION CENTRE

Abergavenny Monk street, Abergavenny. (3254). Also the National Park Information Centre.

Llanfoist Wharf.

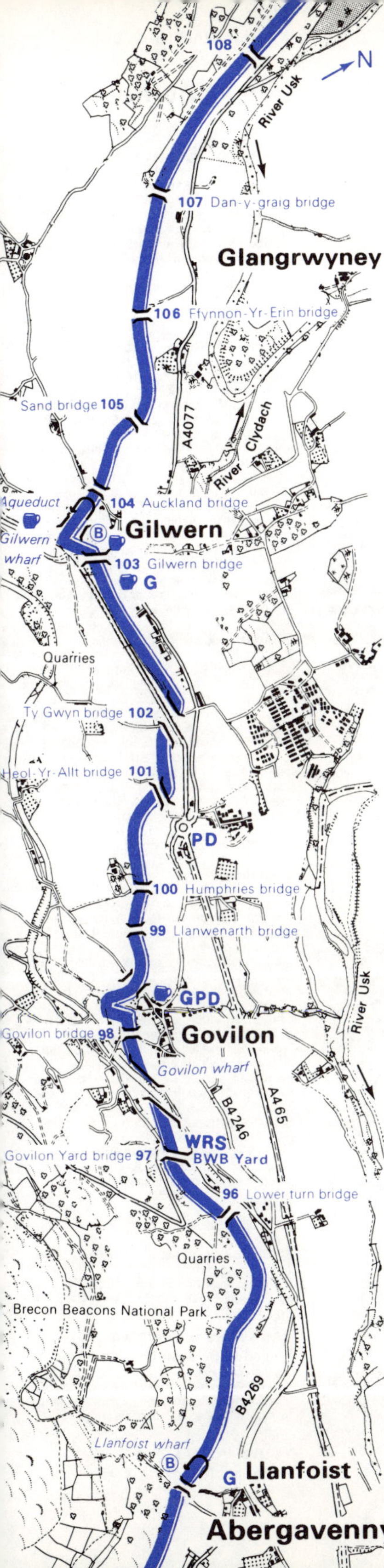

Govilon

5 miles

The canal continues to follow its contour course northwards, cut into the steep, rocky sides of the hill. With Abergavenny spread out in the valley below, and the wooded slopes climbing up from the water, it is a very dramatic stretch. Llanfoist comes into view, partly hidden by the trees. The old wharf buildings, originally built for the tramway that ran up into the hills from the canal, are now a boatyard and hire cruiser base. The boatyard bridge is the best place to leave the canal on foot for Abergavenny; by car, the best road to Abergavenny is from Govilon Wharf. Leaving Llanfoist, the canal continues through the wooded side-cutting. As the course of the long-abandoned railway swings in to join the canal, the towpath changes sides to the west, where it remains until Govilon. The village is huddled beneath the canal; the wharf buildings are now used by Govilon Boat Club, who restored them. The BWB section office is also here. After the big skewed rail bridge (now disused) the canal makes another horseshoe bend, crossing a stream on an aqueduct; there is a picnic site here. As it approaches Gilwern, the presence of the new Heads of the Valley road becomes more obvious, close to the eastern bank. This new road rather shatters the peace of the canal, accompanying it towards Gilwern, having crossed on a big wide new bridge; there are convenient garages near the roundabout, accessible from bridge 101. The canal passes above Gilwern, which spreads down the hill to the east, and then turns sharply before Gilwern wharf. The new road and the hills swing away to the west, leaving the canal to pass through a thickly wooded stretch in comparative quiet; however, the tumbling waters of the river Usk are never far away.

Gilwern
Brecon. PO, tel, stores. The village is built along one main street, which falls steeply away from the canal; there are fine views of the country running down to the river Usk, ½ mile to the east.

Govilon
Mon. PO, tel, stores, garage. The canal passes above the village, and little can be seen from the water. Beside the aqueduct are steps leading down to Govilon. The village is spread out along the road, now quiet after the opening of the new road to the east. It was at one time a small industrial centre, with iron works and lime-kilns, but these have long since vanished.

Llanfoist
Mon. PO, tel, stores, golf course. The boatyard, housed in the old stone wharf buildings, has given the little village a new lease of life. There is a good walk from the boathouse into the mountains, following the course of the old tramway.

BOATYARDS & BWB

Ⓑ **G. N. Price** The Roadhouse, Gilwern, near Abergavenny, Mon. (Gilwern 240). Cafe. Hire cruisers. Water, petrol, gas, dustbins, moorings, winter storage, boat repairs, outboard engine sales and repairs. Small day boats for hire. *Open daily.*

BWB Govilon Yard (Gilwern 328). Water, dustbins, sewage disposal.

Ⓑ **J. R. & J. E. Tod** The Boathouse, Llanfoist, Abergavenny, Mon. (Abergavenny 3877). Hire cruisers. Water, diesel, gas, dustbins, boat building, sales and repairs, inboard and outboard engine sales and repairs. *Open daily.*

PUBS

Beaufort Arms Gilwern. Food, B&B.
Bridgend Gilwern.
Corn Exchange Gilwern. Food.
Navigation Gilwern. Canalside. Fine picture of a navvy and his tools on the sign.
Lion Govilon, below the aqueduct. Food.
Bridge Inn Llanfoist.
Llanfoist Inn Llanfoist.

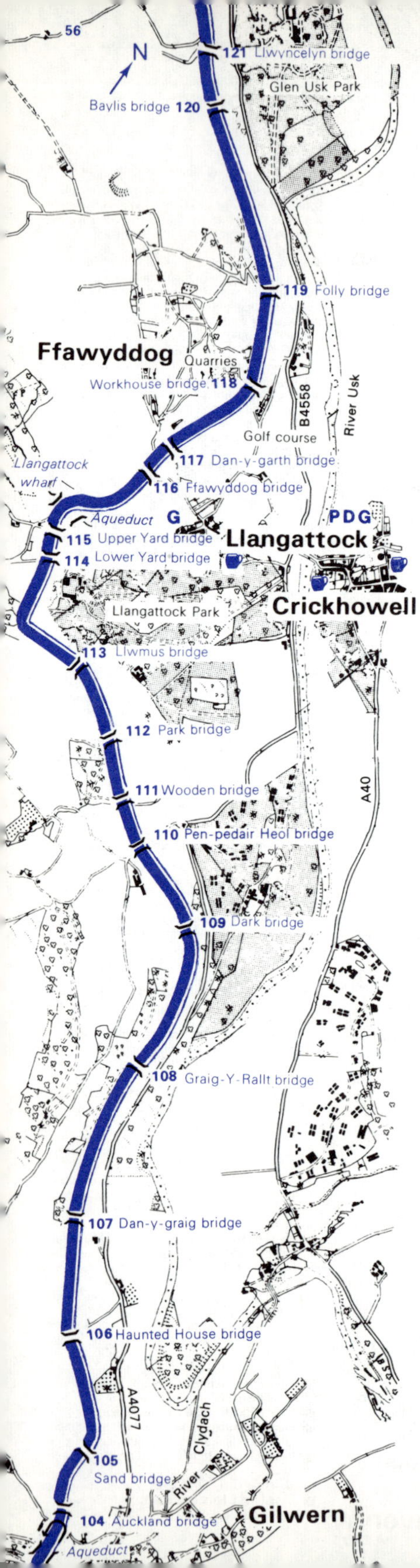

Crickhowell

5 miles

Continuing north west, the canal clings to its contour on the side of the hill which separates it from the river Usk for a while. As the canal passes the old army camp laid out in parkland to the east, the hills become less dramatic. The approach to Llangattock is through flatter country, but trees still surround the canal, hiding the extensive parkland that falls away to the east. Llangattock is set below the canal, best approached from bridges 114 and 115; beyond it lies Crickhowell, the fine houses rising out of the valley over the river. Llangattock wharf is just beyond bridge 115: now a busy mooring site overlooked by the old stone wharf buildings, which include a range of lime kilns. This is a good base for exploring the Brecon Beacons; there are opportunities for caving in the surrounding hills, and horses can be hired in Crickhowell for trekking. Leaving Llangattock, the canal crosses a small aqueduct, and is then quickly back among the hills, whose steep wooded slopes fall sharply to the water's edge. As the river Usk and the canal close together again, there are fine views to the east across the valley. One of the best is across the golf course, which comes right up to the towpath. A short straight then takes the canal to Glen Usk Park, through woods that get progressively thicker.

Llangattock
Brecon. PO, tel, stores. This little village was once famous for its weaving and its lime-kilns. The hills behind the village are riddled with limestone caves and quarries. One cave, Agen Allwedd, has 11 miles of underground passages; the entrance is in the Craig-y-Cilau nature reserve. For access, follow the course of the old tramway west from bridge 114. Permits to enter must be obtained in advance from: Cave Management Committee, 10 Elms Road, Govilon, Mon. Permits for rock climbing, and for specimen collecting, must also be obtained from: Nature Conservancy, Plas Gogerddan, Aberystwyth, Cardigan.

Crickhowell
Brecon. Pop 1,370. EC Wed, MD Thur. PO, tel, stores, garage, bank. The road down through Llangattock leads to the 13-arch medieval stone bridge over the Usk, the imposing entry to Crickhowell. This fine market town, once a centre for the production of Welsh flannel, is spread over the northern slopes of the Usk valley. The town is compact and elegant, with terraces of 18thC and 19thC houses, and some handsome inns. In the centre are the scant remains of the Norman castle. Once controlling a large area, the castle was destroyed during the 15thC, and only the motte and bailey, parts of the curtain wall and a small tower survive. The 14thC parish church contains interesting stained glass. Crickhowell has long been a famous holiday centre, a role it still enjoys today. Apart from hill walking, there are opportunities for fishing and pony trekking.

PUBS

Horse Shoe Llangattock. 20yds from bridge 116.

Bridgend Hotel Crickhowell (338). The nearest pub to the canal, at the end of the mediaeval bridge. Food, B&B.

X **Bear Hotel** Crickhowell (408). Food, B&B.

CARAVANS & CAMPING

Riverside Caravan Park New Road Service Station, Crickhowell (397). 50yds E of Crickhowell bridge. Overnight stops for touring caravans, motor caravans and tents. Caravans for hire, gas, shop. *Open: 1 Mar-1 Nov.*

YOUTH HOSTEL

Ivy Towers, Tower street, Crickhowell, Brecon. (Crickhowell 295).

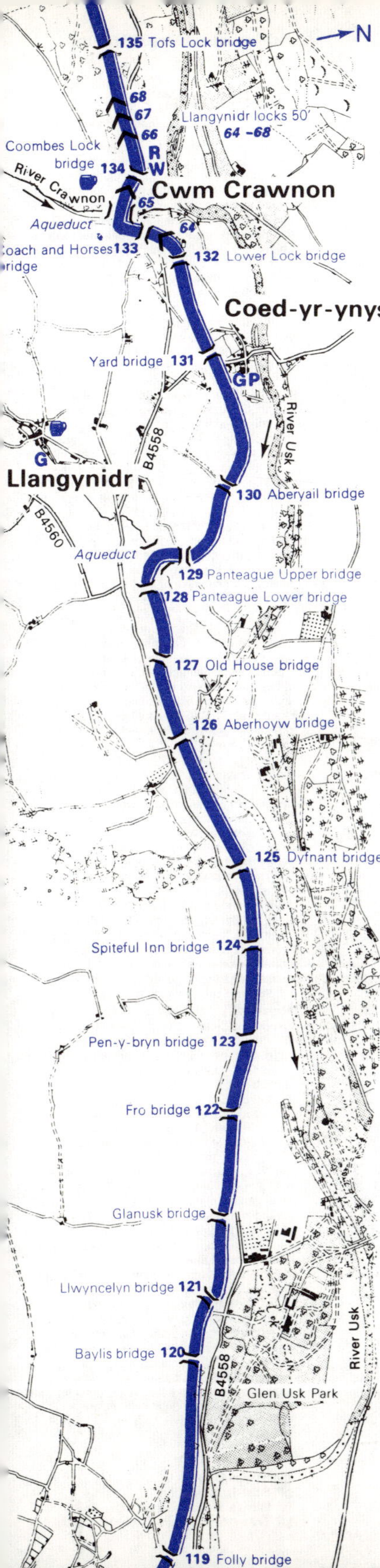

Llangynidr

4¾ miles

Leaving the wooded Glen Usk Park behind, the canal continues its north westerly course along the Usk valley. Among the trees to the east can be seen the Italianate towers of Gliffaes. As the canal approaches Llangynidr, it begins to meander more, leaving the hills and woods behind; river-like, it wanders through open rolling pastureland beside the Usk, passing to the south of the village. Bridge 129 is the best access point for Llangynidr. The small settlement by bridge 131 includes a shop, telephone, tennis courts, and a garage. As this is left behind, the canal reaches the first lock of the Llangynidr flight of five; this, the first lock on the navigable section of the canal, ends the 23 mile-long pound; the short climb to the summit at Brecon starts here. Leaving Cwm Crawnon, where there is a convenient pub and restaurant, the canal turns sharply over an aqueduct and reaches the second lock. On the old wharf beyond the lock is a toll house, now used by BWB; there is a water point, and rubbish bins. Thick woods now flank the canal as the steep hills return on both sides. The Usk valley has narrowed, and so canal and river flow close together through the hills; the canal is still high above the river. In the middle of the woods are the final three locks of the flight, following closely upon each other. The surrounding woods and hills make this the most beautiful setting for any locks anywhere.

Cwm Crawnon
Brecon. Clustered round the canal as it climbs the Llangynidr locks, this hamlet is famous for the Coach and Horses pub, with its French cuisine.

Tretower Court and Castle
Brecon. 2½ miles NW of Crickhowell on A479. Ruins of a late 14thC fortified manor-house, which was built to replace a Norman castle whose remains still exist nearby. The ruined cylindrical keep is unusual.

Llangynidr
Brecon. PO, tel, stores. Spread out over a plateau, this farming village is scattered round the pretty 19thC church. The grandest house in the village is the old rectory, whose grounds run almost to the canal bank.

BOATYARDS & BWB

BWB Llangynidr Maintenance Yard by bridge 134.

PUBS

Coach & Horses Cwm Crawnon, Llangynidr. (Bwlch 245). Canalside. French cuisine.

Red Lion Llangynidr. (Bwlch 223). Food, B&B. Local trout and salmon served in 16thC surroundings. *Good Food Guide 1973. Closed Sun.*

CAMPING & CARAVANS

Castle Farm Upper Llangynidr. (Bwlch 255). 50yds N of lock 67. Overnight stops for touring caravans, motor caravans and tents. *Open: 1 Mar-1 Nov.*

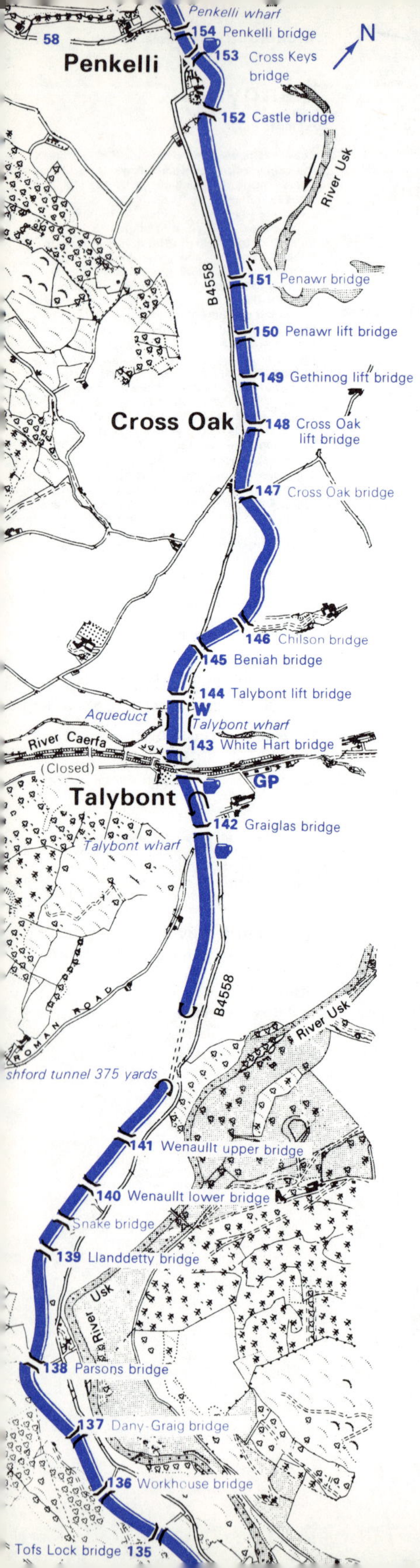

Penkelli wharf
154 Penkelli bridge
Penkelli
153 Cross Keys bridge
N
152 Castle bridge
River Usk
B4558
151 Penawr bridge
150 Penawr lift bridge
149 Gethinog lift bridge
Cross Oak
148 Cross Oak lift bridge
147 Cross Oak bridge
146 Chilson bridge
145 Beniah bridge
144 Talybont lift bridge
Aqueduct
W
Talybont wharf
River Caerfa
143 White Hart bridge
(Closed)
GP
Talybont
142 Graiglas bridge
Talybont wharf
B4558
Roman Road
River Usk
shford tunnel 375 yards
141 Wenaullt upper bridge
140 Wenaullt lower bridge
Snake bridge
139 Llanddetty bridge
River Usk
138 Parsons bridge
137 Dany-Graig bridge
136 Workhouse bridge
Tofs Lock bridge 135

Talybont

5 miles

Leaving the Llangynidr locks behind, the canal continues through thick woods passing the three cottages by bridge 135, a workhouse until recently. The narrow Usk valley pushes the two waterways close together, although the canal stays high above the river. The thick wooded walls of the valley fall steeply on both sides. The minor road that has accompanied the canal all along the valley is still much in evidence, although traffic is luckily very light. The valley starts to open out, and as the hills recede, rolling pastureland flanks the canal as it enters the slight cutting that precedes the short Ashford Tunnel. The tunnel, 375yds long, seems particularly small as it has no portals to speak of; a round hole disappears into the side of a low hill, looking more like a large culvert than a tunnel. The towpath goes over the top, following the line of the tunnel and the B4558. Leaving the tunnel behind, the canal goes straight to Talybont through a low cutting. Passing an old wharf where there is a pub on the road beside the canal, bridge 142, and the disused rail bridge, the canal goes through the village on an embankment, crossing the fast flowing Caerfanell river on an aqueduct. From bridge 142 the remains of a Roman road run to the hill fort above the village. The wharf was used by the Bryn-Oer tramway which brought limestone down from the quarry above. Its course can still be followed. Talybont stretches along the road below the canal, seemingly a typical canal village. At the end of the village is the new lift bridge, the symbol of the restoration of the canal. The bridge is electrically operated; instructions are clearly posted. Leaving the village, the canal enters a different landscape. Pasturelands still roll steeply away to the west, but to the east there are wide flat lands; the canal is carried on a low embankment which continues irregularly for the next three miles. Three newly built conventional lift bridges cross the canal. These are sometimes fixed in the open position to stop livestock crossing the canal; boatmen should always leave them as they find them. A very wide towpath accompanies the canal along this stretch, while on the west side the steep wooded slopes conceal the minor road. A sharp bend takes the canal into the village of Penkelli; the mound of the old castle dominates the village and the canal.

Penkelli
Brecon. PO, tel, stores. This little village was at one time the head of a mediaeval lordship; but the only indication of this today is the castle mound.

Talybont
Brecon. PO, tel, stores, garage. When the railway and canal were both operating commercially, Talybont must have been a busy village. Today it is a quiet holiday centre with facilities for fishing, pony trekking and hill walking, although there is still a busy livestock market. The large wharf overlooks the village, which is clustered round the Caerfanell aqueduct. The river falls rapidly from Talybont reservoir in the hills to the south, to join the Usk.

Llangorse Lake
3½ miles north of Talybont is the largest natural lake in South Wales, 502 acres given over to pleasure and recreation. Boating, yachting, water sports, fishing, pony trekking, caravan parks and camping sites are all available among spectacular scenery. Wildlife is abundant, and the goosander is a frequent visitor. The legendary town of Mara is supposed to lie submerged in the lake; and in fact a crannog, or lake dwelling set on stilts, has been found near the lake's outlet.

PUBS

- **Royal Oak** Penkelli. Canalside.
- **Star Inn** Talybont. Canalside. B&B.
- **White Hart** Talybont. Canalside.
- **Travellers' Rest** Talybont. Canalside.

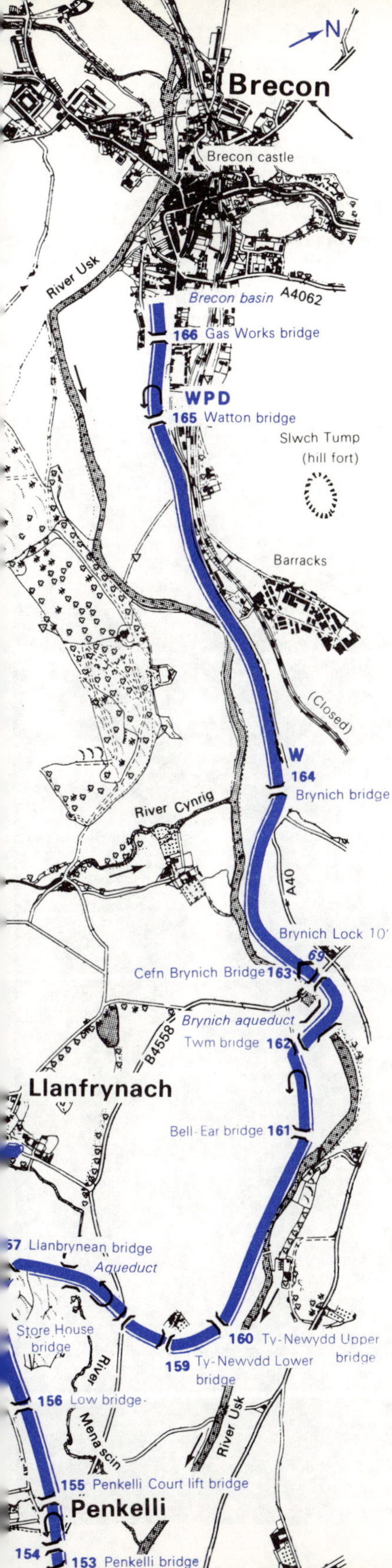

Brecon

4¾ miles

Leaving Penkelli, the canal starts on a long horseshoe bend that carries it through flat wooded country towards the crossing of the Usk. After the last lift bridge, a low embankment carries the canal across marshy ground towards Llanfrynach, but it never goes near the village. The best access point is the B4558 bridge. Before this bridge, a small aqueduct takes the canal over the Nant Menascin; this aqueduct was rebuilt while the canal was closed and so is narrower than the rest. Beyond the aqueduct there is an old mill, now converted to a private house, with a covered loading bay over a small disused arm. The canal completes the long curve back, well above the fast flowing Usk. Llanhamlach lies across the river, approachable via the footbridge over the Usk. In addition to its 13thC church, the area is rich in prehistoric remains. The Usk now stays in sight all the way to Brecon, apart from one small interruption. Bridge 162 takes the towpath to the west bank, where it remains to Brecon, and then the canal turns sharply on to the Brynich aqueduct. This 4-arched stone structure takes the canal across the Usk to the east side of the valley. To the west can be seen the old bridge that takes the B4558 across the river. Immediately beyond the aqueduct is the last lock. Restored in 1970, the lock has a particularly pretty cottage and garden beside it. There is a shop and telephone on the A40. The canal now goes straight to Brecon, high on the hillside, overlooking the Usk all the way. Unfortunately it is also closely accompanied by the A40. Once a Roman road, this now carries a vast amount of fast, heavy traffic. After the peace and seclusion of the whole canal this road comes as something of a shock. The canal is generally below the road, and shielded by a screen of trees, but it is still a rude awakening. Although all the canal is very clean, the stretch into Brecon is remarkably clear and free of weed. The canal follows the road to the outskirts of the town, passing the barracks, and then swings slightly to the west, along the backs of the houses. The entry into Brecon is attractive, with many pretty houses and gardens flanking the canal. Originally the canal went almost to the town centre, turning sharply into a right-angled basin, but now it stops short, at a new basin just beyond bridge 166. The course of the canal to the old terminus can still be traced, although it has vanished beneath a car park and a builder's yard. A big brick warehouse, dated 1892, marks the old head of the navigation.

Brecon
Brecon. Pop 6,300. EC Wed. MD Fri. PO, tel, stores, garage, bank, cinema, swimming pool. Built at the confluence of the Usk and Honddu rivers, Brecon has long been the administrative centre and market town for the Breconshire uplands. It dates back to the Roman period, and although little remains, the narrow streets that surround the castle give an idea of medieval Brecon. Today the town is famous as a touring centre, its cathedral and generous 18thC architecture making it seem more English than Welsh. The Usk waterfront is especially attractive, dominated by the old stone bridge. Sarah Siddons and her brother Charles Kemble lived in the High street.

Brecon Cathedral
Originally the Priory church of St John, founded by Bernard Newmarch, it was given Cathedral status in 1923. Most of the building is 13thC, although the nave is a century later. There is some fine glass, and side chapels dedicated to various medieval trade guilds.

Brecon Castle
Most of the remains of the 11thC castle now stand in the grounds of the Brecon Hotel, and permission to view must be

obtained from the hotel. A large motte and bailey, parts of the walls and two towers survive. The destruction of the castle during the Civil War was hastened by the inhabitants of Brecon, who did not want either side to occupy it.

Brecknock Museum
Glamorgan street. (Brecon 2218). The collections include local history, natural history and a large archaeology section, from pre-Roman to medieval times. The prize exhibit is a dug-out canoe found in Llangorse Lake. *Open: Mon-Sat 10.00–12.00 and Tue-Sat 13.00–16.00.*

Museum of the South Wales Borderers and the Monmouthshire Regiment The Barracks, Brecon. (3111). History of two famous regiments over 280 years. *Open daily 9.00–17.00.*

Llanfrynach
Brecon. PO, tel, stores. An attractive village built in a square round the pretty church. Nearby is the site of a Roman bath house. The white painted pub dates from the 13thC.

PUBS

Many pubs and restaurants in Brecon.

Old Ford Llanhamlach. On A40, ½ mile cross-country from the canal. Food.

White Swan Lanfrynach, ¾ mile from bridges 158 or 157.

YOUTH HOSTEL

Ty'n-y-Caeau, near Brecon. (Llanfrynach 270).

TOURIST INFORMATION CENTRE

Brecon 6 Glamorgan street, Brecon. (2763). Also the National Park Information Centre.

Llangynidr locks. *Derek Pratt.*

Bridgwater & Taunton

Maximum dimensions

The canal is suitable for light craft only.

Mileage

TAUNTON Firepool lock to
Creech St. Michael: 3
Durston: 6
North Newton: $7\frac{1}{4}$
BRIDGWATER Dock: $14\frac{1}{4}$

Total 6 locks

The Bridgwater & Taunton Canal represents a small part of a far more ambitious scheme, the Bristol & Taunton Canal Navigation, for which Rennie gave a quotation in 1811 of no less than £429,990. This was to be part of a ship canal from Bristol to Exeter where it would join up with the long-established Exeter Ship Canal. However, although this sum was forthcoming, very little work seems to have been undertaken and instead, in 1824, an Act of Parliament was obtained to 'abridge, vary, extend and improve the Bristol & Taunton Canal Navigation', which resulted in the much briefer line between the river Parrett at Huntworth (just south of Bridgwater) and Taunton being adopted. The ship canal scheme was abandoned. Although the total cost of the canal on its opening was £71,000, toll receipts in early days averaged £7,000 per year; most of this, however, was drawn from traffic passing to and from the Chard and Grand Western canals which the Bridgwater & Taunton Canal joined at Creech St. Michael and Taunton respectively.

The size of the locks is unorthodox at 54'×13' with a theoretical draught of 3' 0", which meant a normal craft load of 22 tons. The distance as originally constituted from Firepool Lock, Taunton, to Huntworth was $13\frac{1}{2}$ miles. Prior to the building of the Bridgwater & Taunton Canal, there already existed a navigation of sorts on the rivers Parrett and Tone (i.e. an alternative route between Taunton and Bridgwater). This route suffered from drought in summer and floods in winter and so was not particularly reliable; but it was good enough to cause the share-holders of the Bridgwater & Taunton canal great embarrassment, and they were forced to purchase the river navigation in 1832, five years after their own opening.

In 1837 a further Act was obtained authorising the extension from Huntworth to Bridgwater and the building of the dock and its entrance lock from the river Parrett. This led to the curious anomaly of there being two sets of mile stones in close juxtaposition, some of which may still be found. At the time of its jubilant opening on the 25th March 1841, this extension had cost fully £175,000, leaving the proprietors badly out of pocket when the whole waterway and dock complex was sold to the Bristol & Exeter Railway Company for £64,000 in 1866. When control of the waterway eventually passed to the Great Western Railway, little attempt was made to maintain commercial traffic, and the last barge tolls were collected in 1907.

In 1940, at the behest of the War Office, the Bridgwater & Taunton Canal, like the Kennet & Avon, was turned into a line of defence against the possibility of enemy invasion, and pill boxes were erected at strategic points along it. The bridges were fixed and strengthened to carry WD vehicles.

Little interest was shown in the waterway after nationalisation in 1947, although the Bridgwater docks continued to operate under the Railway Executive of the British Transport Commission. In 1958 the Bowes Committee Report on waterways put the canal into category 'C', i.e. suitable for redevelopment, and various surveys were carried out, seemingly without any concrete results. The canal passed to the BWB in 1963, but Bridgwater Docks remained in Railways Board ownership. In the meantime the south western branch of the Inland Waterways Association had begun to consider the canal in terms of restoration, and in 1965 the Bridgwater & Taunton Canal Restoration Group was formed. A year later this group became the Somerset Inland Waterways Society, which was formed to work towards restoration of the canal for amenity purposes. A recent arrangement negotiated by the BWB allows for the extraction of 3 million gallons of water a day from the canal, to provide much needed revenue. At the same time Somerset County Council announced a proposal to raise the bridges to 3'6" headroom, and there is a scheme to convert Bridgwater Docks into a yacht marina.

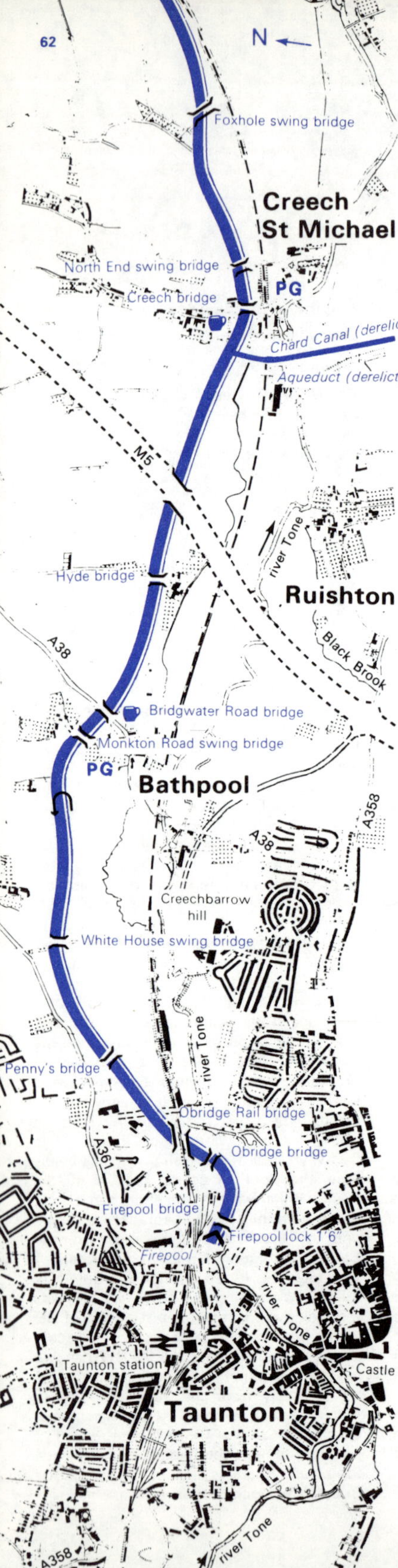

Taunton

5 miles

The Bridgwater & Taunton Canal runs through attractive rolling scenery, typical of rural Somerset. Although not at present navigable, it can be used by small boats and canoes over much of its length; the towpath offers a pleasant day's walk, the most simple way to enjoy the quiet pleasures of the canal. It begins at Firepool lock, at a junction with the river Tone. In the 18thC and 19thC the Tone, which passes through Taunton before meeting the canal, was navigable from Taunton to its junction with the river Parrett. Today sections of the river can be used by small boats and canoes; the Somerset River Authority (West Quay, Bridgwater, Somerset) are responsible for the Tone and also for the Parrett, which is navigable from its estuary on the Bristol Channel to Oath Flood Gates, 30 miles inland. Firepool lock, pleasantly situated beside the Tone weir, is the first stage of the canal's north easterly journey towards Bridgwater, and the estuary of the Parrett. The lock is reached from Taunton by following signs to the cattle market, or by crossing the footbridge over the Tone. The canal curves past the railway station, passing under the Bristol-Exeter line, which follows it closely to Bridgwater. The tall railway warehouse is built on the site of the old junction with the Grand Western Canal. Taunton is quickly left behind as the canal continues through flat pasture land to Bathpool, where the towpath runs briefly through the churchyard of a small corrugated iron church. Before Bathpool is the first of the many low bridges that are the major obstacle to full restoration of the canal. Originally all swing bridges, these were fixed after the swinging mechanism was destroyed by the War Office during the invasion scare of 1940. Contrary to popular opinion, this measure was not so much to prevent navigation by German forces, as to create lines of defence along natural barriers. The Bridgwater & Taunton, like the Kennet & Avon, was hastily turned into a fortified line of resistance. Leaving Bathpool the country becomes hilly and more wooded. A large red brick paper mill and a modern housing estate accompany the canal into Creech St Michael. Before the village is the site of the junction with the Chard Canal. There is little to be seen of the junction itself, but to the south the long embankment crossing the Tone flood plain is still visible, including the stone aqueduct over the river.

Creech St. Michael
Somerset. PO, tel, stores, garage. Although now surrounded by modern housing, the old part of the village still survives. The largely 13thC church is a sturdy and attractive building; inside are fine waggon roofs.

Chard Canal
The Chard Canal was one of the last to be built in England, and was one of the shortest lived of all canals. Work began in 1835, and the $13\frac{1}{2}$ mile line to Chard from Creech St. Michael on the Bridgwater & Taunton Canal was opened in 1842. The canal included three tunnels, two major aqueducts, two locks and four inclined planes. The canal suffered from immediate railway competition, and never made any money. It was finally bought by the Bristol & Exeter Railway Company in 1867 and closed down the next year. Little now survives, but the line of the canal can easily be followed on foot.

Bathpool
Somerset. PO, tel, stores, garage. This small canalside village is now devastated by the passage of the A38.

Grand Western Canal
Originally part of a scheme to link the English and Bristol channels, the Grand Western Canal was to run from Taunton to Exeter, with branches to Tiverton and Cullompton. After much delay, and conflicting advice from many engineers, including Jessop and Rennie, work started

in 1810 on the 11-mile Tiverton branch, which was completed four years later. After another delay, work started on the main line in 1827; the section from Taunton to the Tiverton branch was finally opened in 1838, and included seven vertical lifts and one inclined plane; the site of the plane, at Wellisford, ½ mile north of Thorne St. Margaret, can still be seen. The canal was never completed to Exeter. It was a financial disaster, suffering railway competition almost from its opening day. The main line was closed in 1869, and has largely disappeared. The very attractive Tiverton branch still survives, and is now owned by Devon County Council.

Taunton

Somerset. Pop 37,000. EC Thur. MD Tue, Sat. Taunton has long been a rich agricultural market town, and a route centre for traffic to and from the south west. Although the A38 has made an inevitable impact, the town still has a well-knit and unspoilt appearance. The skyline is dominated by the pinnacled towers of the churches of St. Mary Magdalene and St. James. The first, rebuilt in 1862, is 163ft high; its ornamental splendour rather dwarfs the double aisled church. In Middle street there is an octagonal 18thC Methodist chapel; Wesley preached here when it was opened. The centre of the town is still pleasantly intact, and contains an interesting mixture of buildings; the 15thC municipal buildings, the 18thC Market House, the Victorian Shire Hall and the 20thC County Hall all merge happily together. Remains of the mediaeval town can be seen, including the few fragments of the 13thC priory.

Taunton Castle and County Museum Parts of the building date from the 12thC, but every century has left its mark on the castle. Taunton was a centre for the anti-royalist rebellions of the 17thC. The Great Hall was the scene of one of Judge Jeffreys' 'Bloody Assizes', when 509 rebels were tried. The Somerset County Museum is now housed in the castle. The collections include local and natural history, industrial relics, paintings, a Roman mosaic floor, and fossils from the Mendip Caves. *Open weekdays 9.30–17.30.*

PUBS

Bell Inn Creech St. Michael

New Inn Creech St. Michael

Bathpool Inn Bathpool. Food, camping and caravan site.

Castle Hotel Taunton (2671). Food. Booking essential.

CARAVANS & CAMPING

St. Quentin Guest House Bridgwater road, Bathpool, Taunton (3016). Canalside, by A38 road bridge. Overnight stops for touring caravans, motor caravans and tents. Caravans for hire, gas, showers (h&c), shop. Small boats for hire. *Open all year.*

ne Grand Western Canal at Sampford Peverell. *Derek Pratt.*

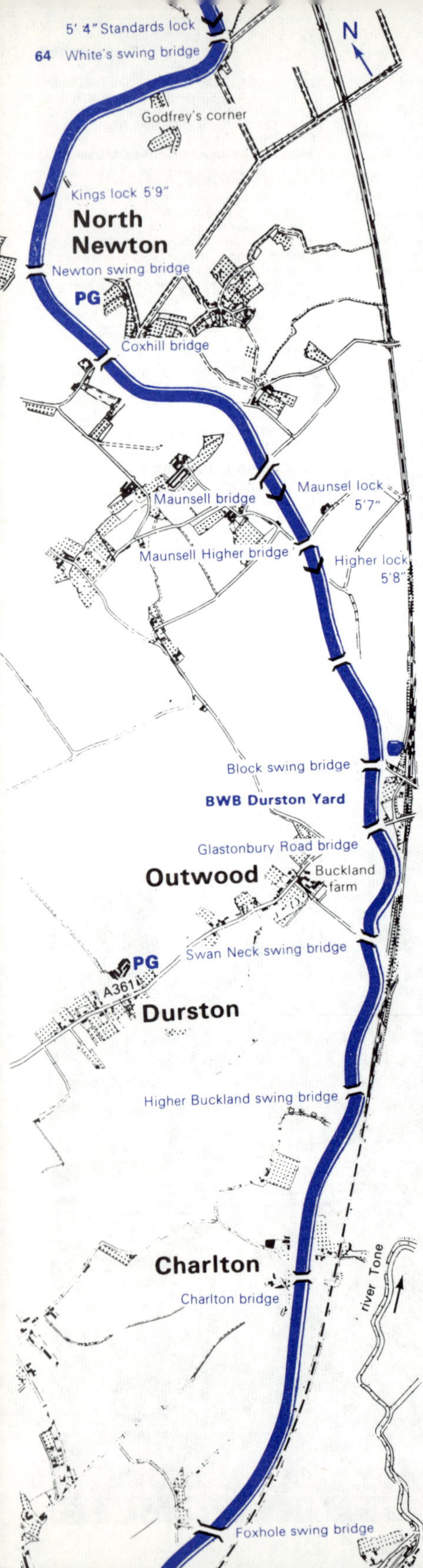

Durston

$5\frac{1}{4}$ miles

Leaving Creech St Michael, the canal starts to swing northwards towards Bridgwater. This section is typical of the quiet, agricultural nature of the whole canal. Arable and pasture lands follow the canal on both banks, while cows and sheep graze by the towpath. The canal is slightly raised above the landscape on a low embankment, and so there are views to the north across the farmlands and to the south across the low-lying flood plain of the river Tone. The canal leaves the river to the south, passing through marsh and moorland. Several small agricultural hamlets flank the canal, revealing again its 'raison d'être'. At Durston, where the BWB Canal Foreman's office is located, the busy A361 crosses. This road, and the nearby railway, are the only interruptions in this very quiet and peaceful stretch. Beyond Durston the canal reaches the first lock, which starts the descent to Bridgwater.

North Newton
Somerset. PO, tel, stores, garage. An irregular farming village, with a pleasing mixture of houses of all periods. The eccentric-looking church was rebuilt first in the 17thC and then again in the late 19thC. Of the original Saxon church nothing remains; but the building is known to have existed because the Alfred Jewel was found here in 1693. This Saxon ornament, the oldest surviving Crown jewel, is now displayed in the Ashmolean Museum, Oxford.

North Petherton
North Petherton is $1\frac{1}{2}$ miles beyond North Newton. It is worth the walk to see the church, whose 15thC tower stands out over the surrounding landscape. 109ft high, it is one of the most richly decorated of all Somerset church towers. The three stages all carry lavish stone ornamentation, and the whole contrasts effectively with the relative simplicity of the church. Inside the light building there is some good Victorian stained glass.

Durston
Somerset. PO, tel, stores, garage. Durston runs in a desultory fashion along the busy A361; although scattered, it is still an attractive village, with a pretty Victorian church. Near the canal, the Railway Hotel stands as a gaunt reminder that there was once a station. Buckland Farm, a grand white house overlooking the canal, stands on the site of an old priory, but nothing remains to be seen of the latter.

BOATYARDS & BWB

BWB Canal Foreman's Office Durston, Somerset.

PUBS

- **Harvest Moon** North Newton.
- **Railway Hotel** Durston.

FISHING THE BRIDGWATER & TAUNTON CANAL

From Taunton through Creech St Michael, Charlton, North Newton, Fordgate and to Bridgwater, there is about twelve miles of excellent fishing. The canal holds most of the coarse fish species throughout its length, and this includes roach, pike, perch, carp, bream and tench. The waterway is weedy in summer, but the fish are of good quality with some of specimen size to be caught. The canal is regularly stocked with the popular species by the local clubs who control the fishing rights.

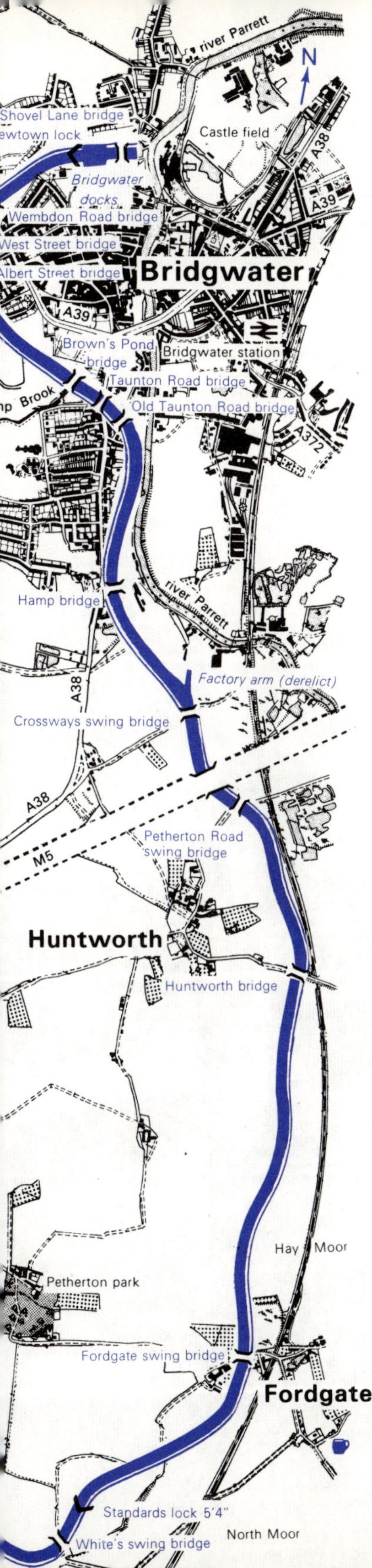

Bridgwater

$5\frac{1}{4}$ miles

Following roughly the contours of the land, the canal continues northwards towards Bridgwater. To the west, low hills rise gradually away from the canal, while to the east low-lying farm and marsh land separates the canal from the river Parrett. This tidal river, a vital drain for the whole area, swings ever nearer to the canal as they approach Bridgwater. To the east the railway line stays close to the canal until just before Bridgwater, when it turns away to pass east of the town. The course of the canal is quiet and relatively isolated, passing only through farms and farm land. There are no villages near the canal, although there is easy access for Fordgate and Huntworth. Standard's lock drops the canal to the Bridgwater level. Nearing Bridgwater civilisation returns with a vengeance, for the M5 Bristol-Exeter motorway has been built over the canal. After the motorway the towpath crosses to the west for the first time, and the canal enters Bridgwater. The canal passes through the town in a cutting, swinging in a wide arc to the west before turning back towards the docks and the junction with the Parrett estuary. The amenity potential of the towpath walk through the town has not been realised, and access is difficult. The canal enters the dock through a stop lock, which passes it into the large inner basin, now sadly deserted. The smaller outer basin is entered through a twin-bascule lift bridge, and then a ship lock and a canal lock used to allow access to the Parrett, and thus to the sea. The river locks have been replaced with concrete barriers, and so access to the canal would not be possible even if it were navigable. Bridgwater dock is owned by the British Railways Board.

Bridgwater
Somerset. Pop 26,700. EC Thur. MD Wed. St Matthew's Fair, held annually *last week in September.* Fireworks Carnival held annually on *Thursday nearest the 5th November.* Bridgwater is an old market town straddling the river Parrett. Formerly the town was an important centre for the cloth trade, which encouraged the development of the port from the middle ages onwards. It was later eclipsed by Bristol, but the old quay still has an attractive 17thC & 18thC flavour, especially in the Dutch gables of the buildings. In the Shipbroker's office there is a Fortin barometer for captains to consult before sailing. Elsewhere in the town are signs of 18thC wealth; the handsome houses in Castle street, built circa 1725, were sponsored by the Duke of Chandos. Little of the mediaeval town survives, largely because of the destructive siege during the Civil War; the 13thC castle disappeared at this time, the water gate on West Quay being the only relic.

St Mary's Church
This 14thC church has been greatly altered over the years, the oldest part being the tower. The most dominant feature is the 175ft spire, added in 1367, which can be seen for miles in every direction. Inside the church there is a fine Jacobean screen.

Admiral Blake Museum
Blake street. This 15thC house is reputedly the birthplace of Robert Blake, Cromwell's admiral. It contains Blake relics and items of local interest. *Open weekdays 10.00–13.00 and 14.00–17.00. (Tue 10.00–13.00 only).*

Bridgwater Docks
Now disused, the docks are a fitting memorial to Bridgwater's erstwhile maritime importance. The entrance locks from the river Parrett have been dammed-off but otherwise the 19thC dock area is still intact. Old winches, bollards, broken cranes and gas lamps surround the stone quays. The old canal and dock office still stands, adorned with brown and white GWR lettering. The huge inner basin, separated from the outer dock by the hand

operated lift bridge, now stands empty and forlorn, overlooked by gaunt 19thC warehouses, and little terraces of cottages. The only boats now to be seen are the dinghies of the sea scouts.

Battle of Sedgemoor, July 6th 1685
When Charles II died, he was succeeded by his brother, James II, who was unpopular because of his Catholic faith. The Duke of Monmouth, an illegitimate son of Charles II, declared himself king in 1685 in Bridgwater. He landed at Lyme with a few supporters and soon raised an army 4,000 strong, including 800 horsemen under Lord Grey. They moved to Bridgwater, while a Royalist army commanded by Lord Feversham and John Churchill (later the Duke of Marlborough) camped near Westonzoyland, three miles east of the town. Monmouth decided on a night attack, but lost the element of surprise when caught crossing the Langmoor Rhine, one of the many drainage ditches in the area. A fierce battle broke out in which the artillery played a dominant part. Grey's cavalry tried to outflank Feversham, but were prevented by the Bussex Rhine, another deep ditch. Although firing continued all night, Monmouth's cause was already lost. The leaders of the rebellion escaped for a while, but the ill-armed rebel army was rounded up, many to be transported or executed on the orders of Judge Jeffreys. To find the site of the battle take the A372 east from Bridgwater; turn left in Westonzoyland. There are signs to the Sedgemoor Memorial Stone.

Fordgate
Somerset. Tel. This farming village straggles in a vague way eastwards from the canal to the deep banks of the river Parrett. It has no centre, and the pub, which overlooks the Parrett, is nearly a mile from the canal. But the walk is not unattractive, typifying the agricultural nature of the area, and of the canal as a whole.

PUBS

Thatchers Arms Fordgate. Overlooking the river Parrett.

CARAVANS & CAMPING

Lakeside Caravan Park 121 Taunton road, Bridgwater (3352). Canalside, just off A38. Overnight stops for touring caravans and motor caravans. Touring caravans for hire, gas, showers (h&c), shop. *Open all year.*

The terminus of the Bridgwater & Taunton Canal. These locks from Bridgwater Dock down into the tidal river Parrett have now been sealed off by the concrete dams.

Gloucester & Sharpness River Severn

GLOUCESTER & SHARPNESS CANAL

Maximum dimensions

Length: 240′
Beam: 30′
Headroom: unlimited

Mileage

SHARPNESS lock to
Purton: $2\frac{1}{4}$
Saul Junction: $8\frac{3}{4}$
GLOUCESTER lock: $16\frac{3}{4}$

Total 2 locks

RIVER SEVERN

Maximum dimensions

Gloucester to Worcester
Length: 135′
Beam: 21′
Headroom: 24′ 6″
Worcester to Stourport
Length: 90′
Beam: 19′
Headroom: 20′

Mileage

GLOUCESTER lock to
Ashleworth: 5
Haw Bridge: $8\frac{1}{4}$
TEWKESBURY junction with river Avon: 13
Upton-upon-Severn: 19
DIGLIS junction with Worcester & Birmingham Canal: 29
Holt Fleet: 36
STOURPORT junction with Staffs & Worcs Canal: 42

Total 5 locks

The river Severn has always been one of the principal navigations in England. Its great length has made it an important trade artery since the medieval period. With its tributary, the Avon, it cuts deep into the heart of England, linking the iron and coal fields with the Bristol channel and the British coastal trade. By using the Severn, boats of a considerable size could sail into the Midlands, and into Wales as far as Welshpool. However the navigation, especially above Worcester, was always difficult, owing to currents, shoals, the demands of water supply for milling etc.

As boats increased in size, and the cargoes became heavier, the navigational problems increased. The larger boats in common use in the 18thC could rarely sail higher than Bewdley, and so by the end of the century this inland port was beginning to lose its significance. At the same time the sandbanks and shifting shoals in the Gloucester area were seriously affecting the trade on the river as a whole. In order for the river to survive as a viable trade route, it became necessary for drastic improvements to be made. Various Acts were passed to ensure the maintenance of the towing path, although the Severn maintained its tradition of using gangs of men to bow-haul boats until well into the 19thC. In 1803 over 150 men were still employed in what Telford called 'this barbarous and expensive slave-like office' on the section between Bewdley and Coalbrookdale.

The demands of increasing navigation, and the spread of canals in the West Midlands (the Staffordshire & Worcestershire Canal linking the Severn with Birmingham and the rest of the network was opened in

1772) led to the passing of an Act in 1793 which authorized a canal to be built from Berkeley Pill to Gloucester. Work began on Gloucester Docks in 1794, and over the next few years 5½ miles of canal were cut. Shortage of money then caused work to be stopped, and so the canal remained useless and incomplete. In 1817 Telford was commissioned by the government to report on the feasibility of the canal, with particular reference to the maintenance of navigation on the Severn. He reported in favour of continuing and completing the canal, but recommended that it should run to Sharpness instead of to Berkeley. The government then put up the money for the canal, mainly to relieve acute problems of unemployment, and after considerable delays the Gloucester & Sharpness Canal was opened throughout in 1827.

Some of the structural problems were caused by the decision to build the canal to ship standard. (At the time of opening, this was the broadest, deepest canal in the world.) But although it greatly increased the cost, this far-sighted decision has ensured that the canal remains in use, and even today Gloucester and Sharpness docks are busy commercial ports.

Once the canal was in use, considerable dredging works and improvements became necessary to maintain the navigation of the upper Severn to Worcester and Stourport. This work, carried out extensively since the formation of the Severn Commission in 1842, included the building of locks and weirs, and the canalisation of parts of the river. The links with the Midlands canal network helped the Severn to flourish, and railway competition increased rather than decreased the traffic both in the docks and on the ship canal. In 1874 Sharpness docks were enlarged and modernised, to handle ships of up to 1000 tons. The same year the Gloucester & Berkeley Canal Company leased the Worcester & Birmingham Canal, to maintain their hold on the trade routes to the Midlands. By 1888 the Severn had a minimum depth of over 6ft, and in most places the depth was 8 to 9ft. Trade continued to thrive, and the recession in the 1920s was soon overcome by the rapidly growing oil traffic, which has been the mainstay of the river until recently. In the last couple of years the level of commercial traffic on the river above Gloucester has fallen away dramatically, but the Ship Canal remains a busy trading waterway.

Spring is a good time to visit the river Severn, when the river banks are speckled white with cow parsley and the riverside alders and willows are bearing their catkin flowers, providing valuable nectar and pollen for early moths and bees. In March and April millions of young eels come upriver with the high spring tides after a three-year, three thousand-mile journey from their breeding grounds in the western Atlantic. The elvers are followed by twaite shad, a sea fish which migrates into fresh water to spawn, and these provide sport for anglers as the fish try to ascend the weirs at Tewkesbury and Gloucester. Other migrants from the sea include 3-foot long sea lampreys and 9-inch river lampreys (which are both parasitic, sucking the blood from other fish), sea trout and salmon. Spring is also the time to listen for nightingales, which sometimes sing in the riverside woodlands, and to watch the caddis flies, may flies and stone flies which have newly emerged from their aquatic larvae and now dance across the surface of the water until they have mated, laid their eggs or fallen prey to some surface-feeding fish such as bleak, rudd or chub. Other fish found in the Severn include bottom-feeders such as bream, carp, roach, gudgeon, loach and occasionally barbel (which was introduced about fifteen years ago), as well as carnivores such as pike, perch and eel.

It is perhaps in the midsummer month of August that the river banks look best with colourful clumps of yellow tansy, purple loosestrife and great willowherb with drooping heads of pink flowers. The recently introduced Himalayan balsam with its handsome pink and white flowers now grows in large clumps at Mythe Bridge near Tewkesbury, Sandhurst near Gloucester and near Gloucester Docks. In marshy areas, riverside pools and withy beds the reedmace grows with brown club-like heads, the common reed with 6 foot high stems and plumed flowerheads, and the greater and lesser pond sedges which bear separate spikes of male (upper) and female flowers. These areas are the home of the sedge, reed and marsh warbler, although the last-named is rare nowadays.

Very few other birds nest along the Severn, although sand martins and kingfishers sometimes excavate nesting holes where the banks are steep, moorhen and mallard are found in the riverside bushes and two heron colonies are known to inhabit tall trees a little way inland from the tidal part of the river below Gloucester. The best place to look for water birds is undoubtedly the area between Frampton and Slimbridge where the extensive mud flats and sand banks exposed at low tide provide an attractive resting place for spring and autumn passage migrants. This area is especially good for winter visitors such as shoveller, pintail, wigeon, teal, lapwing, golden plover, ringed plover, dunlin, curlew, redshank and turnstone, and holds a roosting colony of about 20,000 common gulls. Two species of geese also come to feed on the riverside pastures in winter: up to 5,000 whitefronted geese from Siberia and 100 pinkfooted geese from Greenland.

For the most part, the mammals of our rivers are secretive and nocturnal but water vole holes are a common sight in the river banks and the observant walker can sometimes recognise the tracks or other signs of our rare native otter or the accidentally introduced North American mink, a beautifully sleek black animal a little larger than a stoat, which originally escaped from fur farms but is now breeding in the wild along all parts of the Severn which run through Gloucestershire.

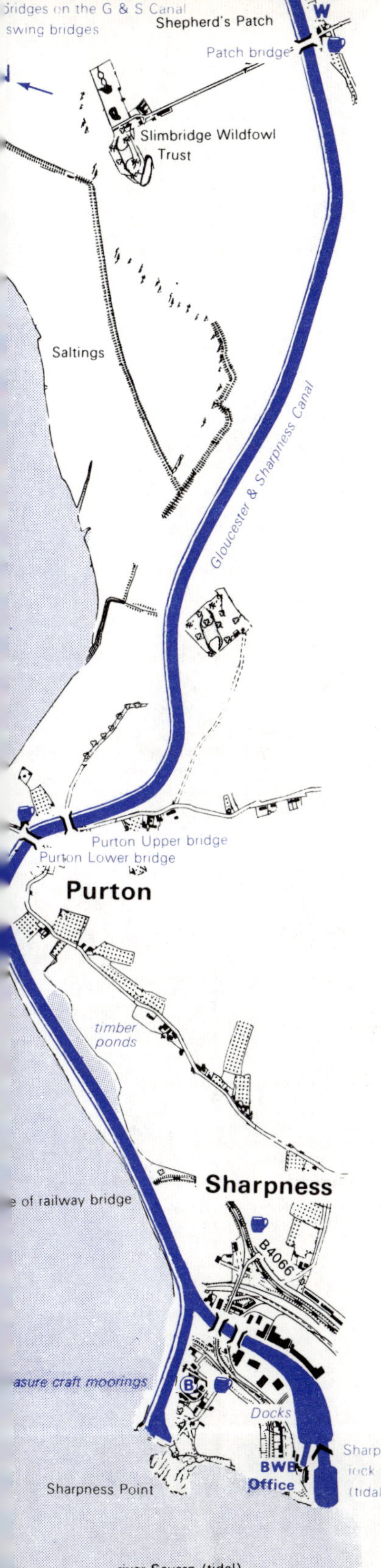

Sharpness

4½ miles

The Gloucester & Sharpness Ship Canal, which was built to bypass the dangerous winding stretch of the tidal river Severn between these two place, has its southern terminus at Sharpness, where there are docks and a large lock up from the Severn. The Gloucester & Sharpness canal is nowadays the only navigable route between the Severn Estuary and the Severn Navigation above Gloucester, so all boats heading upstream must pass through Sharpness lock and docks. It should be noted that the entrance to the lock dries out completely at low water, so boats heading in from the estuary should time their arrival accurately. The best time to arrive at the lock is about 2½ hours before high water when locking down, and about 1 hour before high water when locking up. (The lock is operated only for 2 or 3 hours preceding high water).
It is important to give prior notice of one's intentions to the BWB at Sharpness 228 or 229. Boatmen wishing to proceed down the Severn Estuary from Sharpness are advised not to do so without a pilot. It is important to keep to the marked channel, and the tide runs extremely fast.
Immediately above the lock are Sharpness Docks, which handle ships from all over Europe and Scandinavia. The docks are a very busy area and an important source of revenue to BWB's Freight Services Division. Pleasure boats are encouraged not to dally, but to move on through 2 swing bridges out of the docks and onto the Ship Canal itself. Just north of the 2 swing bridges is an arm off to the west: this leads to a tidal basin which is now, unfortunately, disused. However the length of the arm leading to it is used for permanent and temporary pleasure craft moorings, and there is a small boatyard at the end of it. It is a fascinating place to walk round and see the tidal Severn flowing strongly at the foot of the stone walls. Across the river is the tree-lined west bank of the river, only half a mile across at this point. A railway line runs along the bottom of the hills. However the old 22-arched railway bridge that used to cross the river just north of here has now been completely demolished, and only the merest traces of some of the stone piers can be seen at low water. The bridge was badly damaged one foggy night in November 1959, when a vessel collided with it; the bridge then stood with a hole in the middle until recently, when the rest of it was demolished and the iron girders sold to—of all places—Chile, where they now form a road-carrying viaduct. Along the main line of the canal, the circular stone structure is all that remains of the railway's swing bridge over the canal. A mile from Sharpness Docks, the canal is lined with trees on each side—an uncharacteristically river-like stretch that is belied by the occasional glimpse of the Severn flowing alongside. Old timber ponds open off the canal to the right. Timber was stored afloat here 'in the round'. Then a curve leads to the little village of Purton and its 2 swing bridges. There is only one bridge-keeper here: he spends most of his time at the upper swing bridge. The navigation snakes through the village, passing the big new waterworks before settling down to a steady course of wide, straight reaches. It traverses a quiet, green and predictably flat landscape which is well studded with trees and always bounded to the north by saltings and the mud-flats of the Severn estuary—which is here much wider than at Sharpness. At Patch bridge there are 2 pubs; this is the best access point to the Slimbridge Wildfowl Trust. (*See below*). There is a water point by the bridge.

Shepherd's Patch
This little settlement existed long before the canal: it used to be where the shepherds watched over their flocks grazing the Severn estuary. There are 2 pubs here, a gift shop, cafe and youth hostel.

Slimbridge Wildfowl Trust
Conveniently situated just half a mile NW of the canal at Patch bridge, the Trust should of course be visited. It is close to the river Severn and apart from containing the largest collection of captive wildfowl (160 kinds) in the world, the Trust's grounds and adjacent water meadows attract many thousands of migrant birds–white-fronted geese, Bewick's Swans and all kinds of ducks and waders. Visitors are free to walk all round the Trust's grounds and study the inhabitants, which are fascinating for their variety, quantity and behaviour. Birds with all kinds of exotic markings are seen here–even flamingoes, which are fed a diet of 'red soup' (containing carotin) to maintain their pink plumage. There are plenty of noisy birds here too, the most effective being the screamers and the trumpeter swans. (The latter sound like small elephants).
The Trust does not, of course, exist simply to keep wild birds for the pleasure of visitors. It incorporates an important research establishment that studies all biological aspects of wildfowl, with special reference to ecological trends. Through the ringing of birds here and elsewhere in England, migration patterns are charted and changes in bird populations are calculated. The Trust plays an important role in the defence of the various species from extinction and can already be credited with the rescue of several important species.
Open throughout the year except on Christmas Day: 9.30–17.30 weekdays & Sats, 12.00–17.30 Suns. In Dec, Jan and Feb closing time is 16.30. Telephone enquiries to Cambridge 333.

Purton
Glos. PO, tel, stores. A tiny village of lean, modest houses. It derives an unusual charm from being bisected by the ship canal. The canal is not particularly wide here, and to have a large German coaster quietly throbbing past the cowering post office seems an incredible distortion of scale. The village has, surprisingly, 2 pubs–one is on the canal bank, the other is 100yds away on an enviable site beside the Severn estuary. There used to be a ford for cattle across the river nearby–the herdsman had to judge the time to cross this treacherous river to within a few minutes. Just outside Purton, a huge waterworks has been built for the city of Bristol, at a cost of over £4 million: up to 24 million gallons of water can be drawn daily from the Ship Canal, purified and pumped through a new 4ft. pipe-line down to Bristol for drinking purposes. Small reservoirs have been constructed on the other side of the canal: these will provide a temporary feed if and when a recording device beside the canal nearer Gloucester indicates that the water is too heavily polluted to draw on. The building of this new works explains why one of the swing bridges at Purton is electrically operated and has road traffic lights.

Sharpness
Glos. Tel, bank (Mon, Thur & Fri only). Stores & garage distant. This is an intriguing place. Sharpness exists only for its docks with their tall cranes, old and new warehouses and ever-changing display of foreign ships. It has a strange atmosphere and an interesting situation beside the river Severn. The Severn is wide here, and wild: the tidal range at Sharpness is believed to be the second biggest in the world and the current is very swift, especially when accompanied by the very high winds that often race up the estuary from the sea. Across the water is the hilly Forest of Dean, with a main railway line running almost along the shore. It is only half a mile away, but it could as well be another country, so remote does it seem. In terms of population, Sharpness is very strung out; here and there is a row of terraced cottages, inhabited mainly by dock workers. Along the lockside–the focus of the docks–the few buildings are insubstantial and house only the offices of various shipping firms, HM Customs & Excise and the BWB. And yet the port is a thriving one–in the last 6 years annual tonnage handled here has been multiplied by 6 and is now around half a million tons. Some 20% of this is shipped in containers. Cargoes consist mainly of animal feedstuffs, grain, fertilisers, timber and scrap metal, coming from Ireland, Europe, Russia and Scandinavia. Finnish wooden telegraph poles are also imported at Sharpness. Ships handled here displace up to 5,000 tons (the limiting dimensions of the entrance lock are 55ft. beam by 21ft. 6in. draught). An interesting development at Sharpness is the recent conversion of the old Merchant Navy training camp (on the hill by Sharpness Marine) into an outdoor leisure activity centre for the youth of land-locked Birmingham.

BOATYARDS & BWB

Ⓑ **Sharpness Marine** Floating Yacht Services Store, The Old Dock, Sharpness, Glos. (476). Water, diesel, petrol, gas, chandlery, dustbins. Boat repairs, inboard engine repairs. Coffee & snacks served. Groceries for sale (the most convenient place in Sharpness for boatmen to buy victuals). Sewage disposal and slipway available in 1973, also rowing boats to hire. *Open daily.*
BWB Sharpness Office beside Sharpness lock. Maintenance depot (tel. 348) and freight services office (tel. 228).

PUBS

Patch Hotel, opposite the Tudor Arms.
✕ **Tudor Arms** Shepherd's Patch, a few yards from Patch Bridge. Free house; restaurant and bar snacks.
Berkeley Hunt, Purton. Free house, by the lower bridge.
Berkeley Arms, Purton. Free house, 150 yards from the lower bridge. Situated beside the river with an excellent view along it.
Severn Bridge & Railway Sharpness, on a hill to the north east of the docks. The pub sign shows a picture of the former railway bridge.
Sharpness Hotel Sharpness, on a hill to the north-west of the docks. A large building on the hill near the Sharpness Marine boatyard, this is not so much a pub as a remarkable no-holds-barred night-club for rockers.

YOUTH HOSTEL

There is an hostel at Shepherd's Patch, near the canal. (Cambridge 275).

Unusual visitor at Sharpness: a Thames Sailing Barge on its way up to Gloucester.

Saul Junction

$4\frac{3}{4}$ miles

The Ship Canal continues towards Gloucester, with the spacious Severn estuary over to the west. In contrast to the river Severn above Gloucester, one can easily see out over the landscape, which around here is perfectly flat but quiet and green. Swing bridges punctuate the canal and at almost every bridge is a bridge keeper's cottage. These cottages are peculiar to the Gloucester & Sharpness Canal and have great charm—they are only small single-storey buildings, but each one has a substantial classical facade with fluted Doric columns and a pediment; these features give the houses a uniform dignity. At Cam Bridge is an unnavigable arm which feeds the canal with water from the Cotswolds. At Frampton on Severn the church is passed on the east side, then after a long straight the navigation bends to the north-east. Trees encroach here on one side, a chocolate factory appears on the other, and several bridges lead past scattered houses to Saul Junction, where the canal resumes its former straight and lonely course through the flat landscape. The site of a former bridge is evident from the temporary narrowing of the channel. Over to the east the great Cotswold ridge marches parallel to the canal.

Saul Junction
This unusual waterway 'crossroad' is where the Stroudwater Canal intersects the Gloucester & Sharpness Canal. The former canal was an extension of the Thames & Severn Canal, which used to run from the Thames at Lechlade, through the Cotswolds via the great Sapperton Tunnel and thus to Stroud. 12 locks brought the Stroudwater Canal down from this point to Saul, where it crossed the Ship Canal and continued to Framilode. Here it locked down into the tidal river Severn. Like the Thames & Severn, the Stroudwater Canal has been disused for many years, and the old lock by Saul Junction is derelict. However a 300 yard stretch to the south-east from the junction to the first bridge (now lowered) is navigable and in use as a mooring site. At the junction itself is a swing bridge and a cottage (previously the 'Junction Inn') as well as plenty of boats at the private repair yard. The village of Saul is half a mile to the west. It is worth walking the mile from Saul Junction to the river Severn. The towpath is in good shape and this isolated section of the Stroudwater Canal is still in water. There is a pub (the Ship) along the way and from the riverside church in Framilode a footpath runs beside the Severn to the Darell Arms, whose gardens overlook the river.

Frampton on Severn
Glos. PO, tel, stores, garage. A beautiful linear village notable mainly for its green, which is about 100 yards wide and fully half a mile long. All manner of attractive houses attach themselves to the edge of this magnificent expanse of common land, and the occasional cars that drive down the middle of it are kept in their proper scale. Trees and ponds are scattered along it; the gateway that guards the Court is on the east side. The church of St. Mary is at the south end of the village, near the canal; it is mainly of the 14thC. The stained glass and monuments are worth a look, and the Romanesque lead font is one of only six in Gloucestershire.

Frampton Court
Facing the village green is this Georgian mansion (1731–33) whose gardens contain a Gothic Orangery (1745) and an octagonal dovecote. *Visits by written application only, to Mrs Clifford, Frampton Court, Frampton on Severn, Glos.*

PUBS

Bell hotel Frampton on Severn, $\frac{1}{4}$ mile SE of Fretherne bridge.

Three Horseshoes Frampton, halfway along the green.

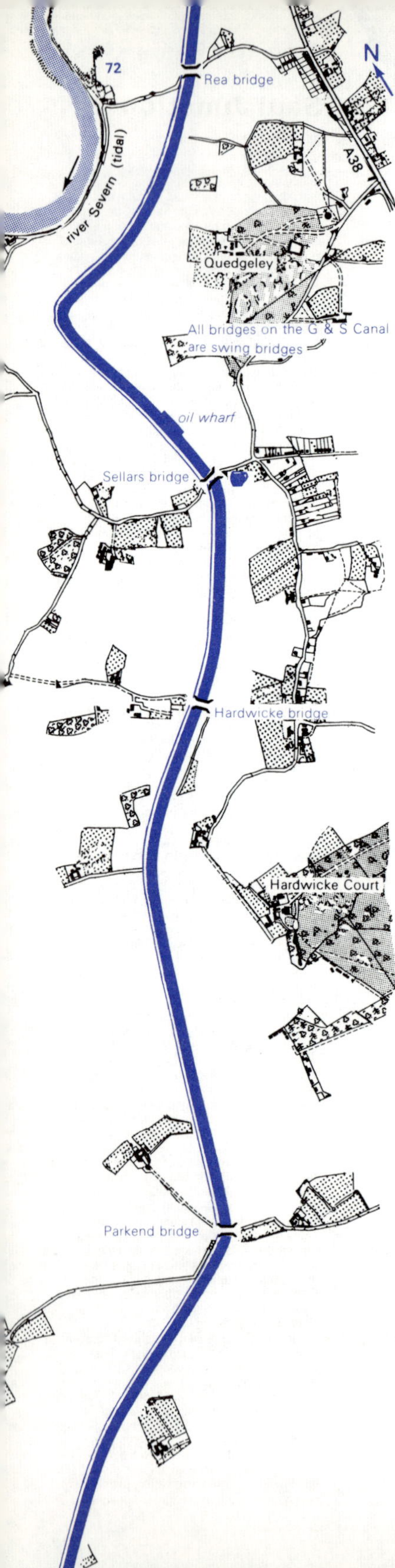

Hardwicke

$4\frac{3}{4}$ miles

On from Saul, the canal continues through the undramatic country, which is slightly wooded. There are no villages on this section, but several farms are situated near the canal. Towards Sellars Bridge, the navigation enters a cutting for the first time since Sharpness. There is a pub by this bridge, and just to the north is an oil wharf for small ships. This is the furthest (northernmost) point which the oil traffic reaches on the Severn Navigation – oil used to be carried further on past Gloucester and right up to Stourport in barges, but this is all finished now. North of Quedgeley wharf the canal approaches the river Severn (hidden behind a flood bank and far narrower up here than downstream); turning sharply eastward, the canal reaches Rea Bridge.

Navigational Note

The 3 bridges Sellars, Rea and Sims (*see next page*) have a greater headroom (over 7ft.) than the others on the Gloucester & Sharpness Canal. Boats normally used on the narrow canals will find no difficulty in getting under these 3 bridges without them being opened, although they do so at their own risk. Boats should not pass under these bridges at night without receiving a green light from the keeper.

The Severn Bore

This famous natural phenomenon occurs on that section of the river which is bypassed by the Gloucester & Sharpness Canal. The Bore is a wave that travels upstream: it is created by the strong tidal flow encountering the 'land water' and chasing it back up the shallow, winding river. One of the best places from which to see the Severn Bore is Stonebench, Elmore – only 500 yards west of Lower Rea bridge on the G & S Canal. (Another good place is Maisemore Weir, above Gloucester.) A nominal Bore occurs on several consecutive tides each month, but a substantial wave of over 7 to 9 feet is much rarer. Local enquiry will indicate forthcoming dates. *See 'The Severn Bore' by F. W. Rowbotham.*

PUBS

Pilot Canalside, at Sellars bridge.
Castle Guest House Canalside, at Park End bridge. Lunch and dinner daily throughout the year. B&B. Groceries available, also diesel fuel for boats. Enquiries and reservations to Hardwicke 328.
Anchor Epney (on the river $1\frac{1}{2}$ miles west of Park End bridge).

FISHING THE GLOUCESTER & SHARPNESS CANAL

This canal used to be disastrous fishing until a few years ago, but it has now improved dramatically and is now a popular fishing waterway. In addition to the usual species, dace, roach and chub are particularly plentiful. Bream can be taken at places like Saul Junction (near Frampton) and Hempsted. Being a ship canal, this waterway is of course relatively deep – up to 14ft. – with steeply shelving sides. The fish tend to keep to the margins, especially where there are plenty of weeds, so it is best for the determined angler to keep away from the lengths which have been recently piled. Annual permits or day tickets for this canal are available from the British Waterways Board, Dock Office, Gloucester.

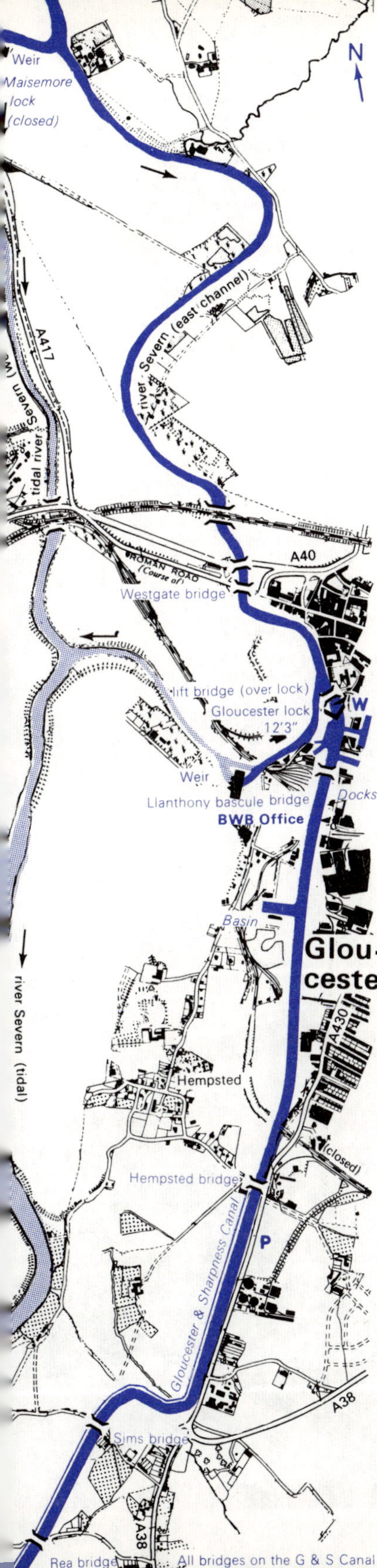

Gloucester

5½ miles

At Rea Bridge the canal enters a cutting and describes a sharp double bend, from which one emerges into a completely different landscape: the quiet countryside has disappeared, its place taken by outlying industrial works on either side of the main road that runs noisily parallel to the navigation. The Ship Canal, strangely enough, plays little part in the generation of wealth that this industry represents, but north of Hempsted Bridge is a large timber wharf for discharging ships bringing imported wood. Further on is the oil dock, a grain silo and a general cargo quay. Ahead is Gloucester, and the extensive docks that are laid out virtually in the town centre. These are superb docks, for all around are the great warehouses ranged along the quays. Boatmen wishing to moor here – the best place for visiting Gloucester – should go to the office by the lock and seek advice from the BWB official. Gloucester lock marks the northern end of the Gloucester & Sharpness Canal, and lowers boats back into the river Severn, reminding one that the Ship Canal is well above the river level. It has to be kept filled with water by pumping up from the river. No boats at the tail of Gloucester lock should follow the river to the south-west, for Llanthony lock is closed and only a weir awaits them. North of Gloucester lock, the river is bounded on the town side by a high quay, with moorings more suitable for large vessels than for motor cruisers. Gloucester gaol is nearby. Proceeding upstream, boatmen will find themselves on a dull length of river, narrow and hemmed in by high banks. A sharp bend requires a careful look-out; beyond it are road and rail bridges. The river winds along in its own isolated way, flanked mainly by trees. At one point it approaches a minor road, but the former pub here is now a private house. Further upstream is the junction with the big western channel of the Severn, whose separate course between here and Gloucester explains the narrowness of the navigation channel. There is in fact a lock (Maisemore lock, now closed) just 300yds along the western channel of the Severn. This is a relic of the days before the Ship Canal was built: it used to give access from the upper Severn to the Herefordshire & Gloucestershire Canal (now derelict) which joined the Severn near Gloucester.

Gloucester
Pop 90,000. MD: Street Market Sat, Cattle Market Mon & Thur. 2 stations. Now a busy manufacturing town, commercial centre and port, Gloucester was once the Roman colony of Glevum. The town was laid out in a cross plan, with north, south, east and west gates. This geography still survives, if only in name. Traces of Roman habitation are much more difficult to find than in other Roman towns in Britain, but when the Bell Hotel was being demolished in recent years, excavations revealed 1,000 square feet of paved courtyard, believed to be the site of the Roman forum. There are a few interesting old buildings in the town centre, notably the 12thC Fleece hotel and numbers 11–15 Southgate, but otherwise the town centre is of less interest than one might expect. However the glorious Cathedral provides an oasis of peace and beauty in the town. The other area of real interest is the docks.
Gloucester Cathedral Founded as an abbey in 1089 by Abbot Serlo, this splendid building is essentially Norman, but extensive remodelling of the choir and transepts between 1330 and 1370 shows fine examples of early Perpendicular architecture. These alterations were authorised by King Edward III, whose father Edward II was murdered at Berkeley (2 miles south of Sharpness). Gloucester Cathedral was the only place that would offer him burial in consecrated ground, and his tomb here became the object of

pilgrimage. The new king showed his gratitude to the then Abbot of Gloucester with financial assistance. The great east window is particularly fine—the glass dates almost entirely from 1350, when the window was built, and is one of the first examples of a church window depicting rows of people. But perhaps the most interesting part of the present building is the adjacent cloisters: they feature the earliest known fan-vaulting (mid-14thC). It is still in good condition and is delightfully ornate.

City Museum & Art Gallery Brunswick rd. (Gloucester 24131). Exhibits of furniture, glass, silver, costumes and coins; also local archaeology, geology and natural history. *Open weekdays 10.00–17.30.*

Folklife and Regimental Museum 99–103 Westgate st. (Gloucester 24141). Housed in 3 Tudor timber-framed buildings known as Bishop Hooper's Lodging and scheduled as an ancient monument. Fine collection of local history and bygones: there is a whole section on the river Severn, its vessels and the salmon and eel fishing industries that it once supported. Relics of the siege of Gloucester (1643) are here, also the collection of the Gloucester Regiment. *Open weekdays 10.00–17.30.*

Gloucester Docks These extensive docks close to the centre of Gloucester are, to many people, really much more interesting than the town itself. They are at the north end of the Gloucester & Sharpness Ship Canal, where it locks down into the Severn. The docks, which were built around 1827, accommodate vessels up to 700 tons, handle less oil traffic than they used to, but more 'dry' cargo. Imported timber and grain are 2 of the main cargoes brought up here—they arrive mostly in big barges from ports down the Bristol Channel, for nowadays only a few coasters a week navigate the length of the canal. One may be sure that if the canal had not the generous dimensions that it has, Gloucester Docks would be disused by now, like the wharves up-river at Worcester and Stourport. The 7-storey dock warehouses are magnificent. Fully 9 of the original buildings still stand, lining the waterside like block houses and concealing it from the town. The docks have now gained an interesting new feature, for the old Llanthony swing bridge carrying a road across the docks has been recently replaced by an elegant steel bascule bridge, which carries heavier traffic and is faster to operate than the old one. Gloucester Docks add up to a fascinating scene, and yet one that offers plenty of scope for improvement. There is as yet no boatyard for pleasure boats (the nearest one is down at Sharpness); perhaps as the old warehouses at the docks become too outdated for modern storage, one could be converted into a chandlery, store or workshops for boats.

BOATYARDS & BWB

BWB Gloucester Area Engineer's Office The Dock Office, Gloucester (25524). Situated outside the docks not far from Gloucester lock. This number should be contacted by all boatmen intending to navigate the Ship Canal at weekends. Water point available beside the lock.

FISHING

Local anglers regularly catch chub, roach and dace in the Gloucester area, but the best fishing here is for the bream.

🛈 TOURIST INFORMATION CENTRE

Gloucester The Guildhall, Gloucester. (22232).

Gloucester Docks.

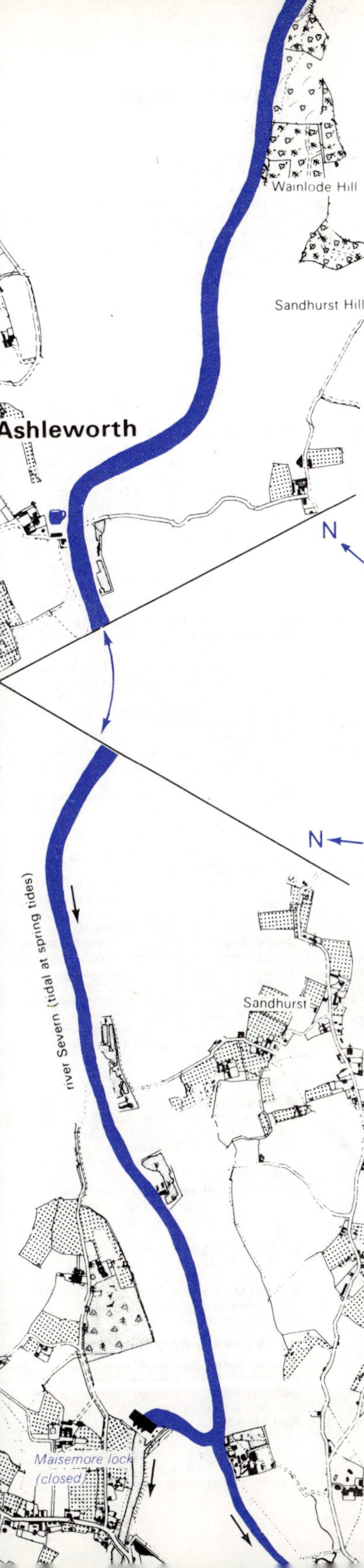

Ashleworth

$4\frac{1}{2}$ miles

Leaving the junction (known as the 'Upper Parting') of the east and west channels of the river Severn, the river from this point northwards is predictably wider. Its character changes very little in all its journey to Stourport—most of the way it is lined by trees and high banks. The surrounding countryside that accompanies the river is quite pretty but because of the banks, the boatman will see little beyond except for the occasional hills. The walker along the banks is luckier, in having good views of the river and the surrounding countryside; but the towpath has mostly been eroded away. Away from the centre of the river it is often extremely shallow; there are anyway few mooring places, so access to the villages on either side is severely limited. After a series of long reaches, the spire of Ashleworth church appears on the left side as the river bends to the east towards the hills that rise steeply from the river-bank. Unfortunately access to Ashleworth from the river is poor. The main hill here is Wainlode Hill, which reaches a height of almost 300ft.

Navigational note
Because of Wainlode Hill's susceptibility to erosion by the river, old barges have been sunk in the river near the east bank of the section on this page, in order to protect the river bank. All boats should keep to the *west* side of the river to avoid the hulks. The area is marked by posts.

Ashleworth Quay
Glos. Behind the tiny isolated pub is a fascinating group of 15thC buildings, all virtually intact. Ashleworth Court, a long, low, stone building, was completed in 1460 and stands next to the church. *(Open only to parties by appointment to Mr. H. J. Chamberlayne, Ashleworth Court, Gloucester*—telephone Hartpury 241). The church, with its pretty spire, is nearby; but of greater interest is the big stone tithe barn (125ft. by 25ft.). This is owned by the National Trust but is still used as a working barn by the farmer—open daily during daylight hours. Ashleworth Manor and the rest of the village are set well back from the river. The Manor is contemporary with the Court, and is of timber-framed construction. *It is open only by written appointment to Dr Jeremy Barnes, Ashleworth Manor, Ashleworth, Glos.* There used of course to be a jetty at Ashleworth Quay; but this has vanished now, and boat crews wishing to get ashore must either ground their craft and wade ashore through the mud to the bank or tie up to an overhanging tree where there is enough water (a few yards downstream of the pub).

PUBS

Boat Ashleworth. A delightful isolated pub on the river. Access is difficult (shallow water).

FISHING THE RIVER SEVERN

The Severn is a very good waterway for fishing, and predictably better than most canals. It is famous for its eels, which as elvers run up the river in huge shoals every year. The Severn has always been famous for its salmon, and these are still to be found. Other common species to be found in the river include roach, dace, bleak, gudgeon, perch and pike. The weir pools that are close to every lock are the best places to fish. Barbel, which were introduced to the Severn a few years ago, have thrived, and multiplied greatly. Nowadays large catches of these powerful fighting fish are to be caught. In the Tewkesbury, Upton and Worcester reaches these fish are present in large numbers, as are chub.

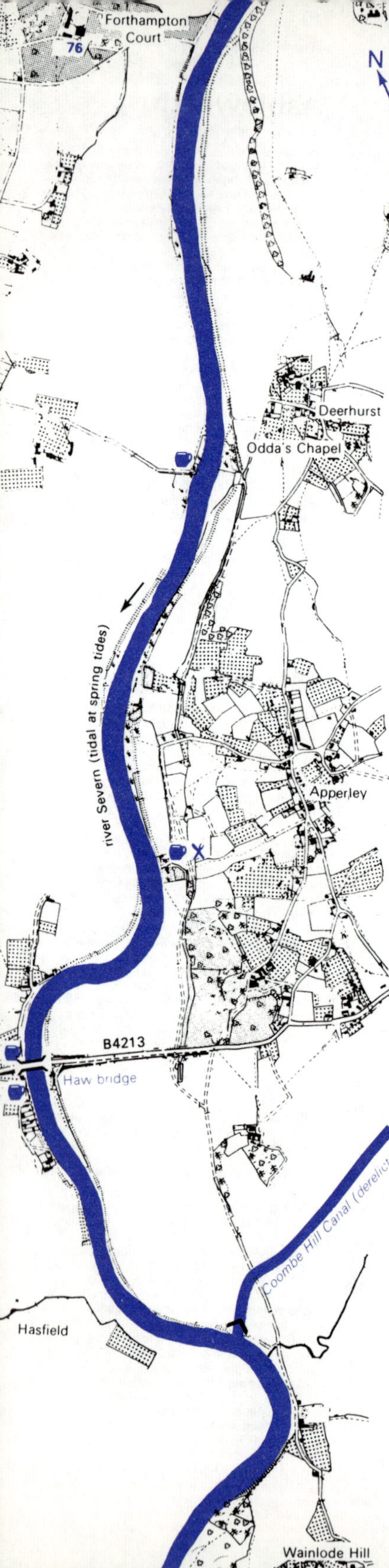

Haw Bridge

5 miles

The towering mass of Wainlode Hill slowly recedes as the tree-lined river curves round to the north-west. The silted up lock on the east bank is the entrance to the former Coombe Hill Canal, now completely derelict. The modern bridge to the north is Haw Bridge. There are pubs and good moorings. Navigators should keep away from the east bank near this bridge—there is a submerged obstruction. The river winds through an S-bend, passing a riverside pub and a line of hills to the east; then it straightens out somewhat as it heads for Tewkesbury. Yet another riverside pub, the Yew Tree Inn, is passed—a half-sunken barge serves as a mooring. A sailing club is based here. Opposite is Odda's chapel, but access is bad because of the rocky banks.

Deerhurst
Glos. Tel, stores. The most important feature of this small village is the beautiful church of St. Mary, parts of which date back to the year 804. The font with its trumpet-spiral motif dates from the late 9thC and for some time the bowl had been used as a wash tub in a farm. It was discovered and reunited with the stem in the late 19thC, and is now thought to be one of the best preserved Saxon fonts in England. The church contains some interesting brasses and 15thC stained glass. Access is bad from the river, as the banks are rocky.
Odda's chapel about 200yds SW of the church. This Anglo-Saxon chapel dates from 1056; for years part of a farmhouse, it was rediscovered for what it is during repairs in 1885. It is now classified as an ancient monument and *open at any reasonable time.*
Haw Bridge
The old cast iron bridge, built in 1824, was knocked down accidentally by a barge in December 1958, when the river was in spate. The replacement bridge was opened in 1961, a few yards downstream of the old one. There are two pubs at the bridge, but little else. The villages of Tirley and Hasfield are a mile to the north west.
Coombe Hill Canal
This short canal, running eastwards from the Severn for 2¾ miles, was built in 1796 to facilitate the carriage of coal to Cheltenham. There were two locks. Despite high tolls, the canal never made any money, and changed hands several times during the 19thC. In 1875 the lock gates were swept away by floods, and the canal was formally abandoned the following year. Parts of the canal are still in water, and so its course can easily be traced today; the old warehouse at Coombe Hill basin still stands.

PUBS

Yew Tree Inn on west bank of river, opposite Deerhurst. Large pub at the end of a lane. Sailing club based here; food always available. Reasonable temporary moorings. The quiet village of Chaceley (*PO, tel.*) is a mile up the lane.
Coal House on east bank of river near Apperley. Moorings not brilliant, but food is available, also a skittle alley. (Tirley 211).
New Bridge Haw Bridge.
Haw Bridge Inn at Haw Bridge. Moorings. Skittle room.
Red Lion on east bank of river beside Wainlode Hill. Popular with motoring families, but no moorings.

CARAVANS & CAMPING

Severn Side Apperley. (Tirley 276). Riverside, S of Haw Bridge. Overnight stops approved for touring caravans, motor caravans and tents. *Open Mar-Oct.*
Red Lion Wainlode Hill, Norton. (Twigworth 251). Riverside. Overnight stops approved for touring caravans, motor caravans and tents. Caravans for hire, gas, showers (h&c), shop. *Open all year.*

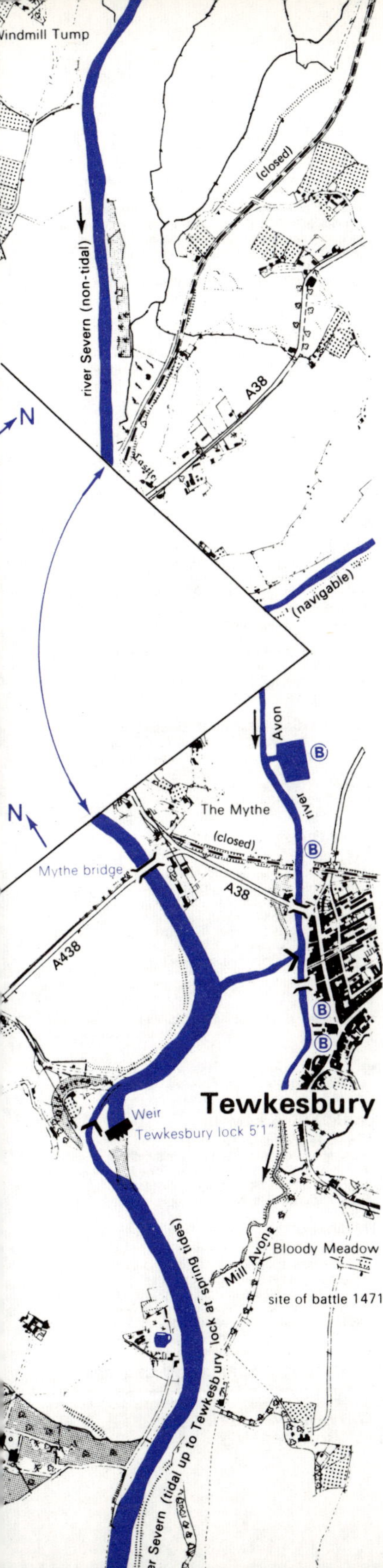

Tewkesbury

3½ miles

The river now passes another pub, facing a sailing club. There is a private ferry between them; access to the pub is difficult because of shallow water. One of the 2 channels of the Warwickshire Avon enters here from Tewkesbury: the battle of Tewkesbury was fought just to the east of here in 1471. Marked by the Abbey, Tewkesbury can be seen to the north-east, but the Severn sweeps round to the west of the town, leaving an enormous expanse of flat, empty meadow between them. The big lock is well concealed on a corner between the weir and a backwater. (The weir is, incidentally, the highest point to which normal spring tides flow). Upstream of the lock is a junction with the main (navigable) course of the river Avon—boats heading for the Lower and Upper Avon Navigations should turn east here, as should boats intending to visit Tewkesbury. Beware of the shallow spit projecting south-west from the tip of the junction. Continuing up the Severn, one reaches the single 170ft. span of the cast iron Mythe bridge over the river, built by Thomas Telford in 1828. Steep wooded hills rise on the east bank by this bridge, but the river bears off to the north west and soon leaves them behind. Its character remains virtually unchanged—it is lined by high banks and trees, untouched by villages or towns, and seemingly isolated from the countryside that its wide course divides so effectively. There is, unfortunately, little commercial traffic on this stretch: two or three motor barges a week still carry grain up to the big mill on the Avon at Tewkesbury, but all other regular trade on the river upstream of here has now ceased, and apart from the passage of the occasional pleasure boat, the waterway seems curiously lifeless.

The river Avon in Tewkesbury
It is certainly worth turning off the Severn into the river Avon—this is the way to Tewkesbury, Evesham and now to Stratford-on-Avon. Boaters not wishing to buy the short-term pass on to the Lower Avon Navigation may tie up just below the big Healing's Mill to visit Tewkesbury, but those who decide to go through the pretty Avon lock (operated by a lock keeper from 9.00a.m. to 9.00p.m. daily) will find it a worthwhile diversion. The Avon is a tremendous contrast to the river Severn: it is much smaller and prettier, and its low banks mean that one's view of the delightful surrounding countryside is never restricted. There are some splendidly situated locks and several magnificent watermills on the river. It is deservedly a very popular navigation, and the large numbers of motor cruisers and sailing boats add a lively dimension to the landscape. The Avon in Tewkesbury is most charming, full of moored boats. (There are several boatyards, and a new marina). Upstream is the ancient King John's bridge, downstream a branch leads to the 12thC Abbey Mill, a superb white clapboarded watermill. Big sluices here (one of which used to have a self-adjusting gate) mark the terminus of the branch. There are some fine black and white timbered cottages nearby, looking across the meadow to the Severn.

Tewkesbury
Glos. Pop 8,500. EC Thur. MD Wed, Sat. All services. An historic town at the junction of the rivers Avon and Severn, Tewkesbury is well worth visiting. There are many attractive and ancient buildings to see, chief among these being of course the Abbey. One of the more unusual aspects of Tewkesbury is the great number of tiny alleys leading off the main street that is the backbone of the town. These alleys yield tempting views of discreet cottages, gardens, back walls and private yards. One of these—Baptist Chapel Court—leads to the old chapel, built around 1655. This tiny, simple building and its little burial ground

reflects well the modest aspirations of the minority Baptist movement. Believers used to come from Cheltenham (8 miles east) to attend meetings at this, one of the oldest Baptist chapels in England. There are very many other buildings of great interest throughout Tewkesbury, chiefly of the timbered variety, with overhanging gables. Some have curious names like 'House of the Nodding Gables' and 'Ancient Drudge'. There is also a liberal scattering of historic pubs, notably the Hop Pole Inn (associated with Dickens' Pickwick Papers) and the Bell hotel, an Elizabethan building which was Abel Fletcher's home in the book 'John Halifax, Gentleman'.

Tewkesbury Abbey This superb building is cathedral-like in proportions and is generally reckoned to be one of the finest Norman churches in the country. It is contemporary with Gloucester Cathedral, and has the same type of vast cylindrical arches the length of the nave. This massive scale is repeated throughout the building: the beautifully decorated central tower, 46ft. square and over 130ft. high, is the largest Norman tower in existence. The recessed arch that frames the mighty west window is over 60ft. high. The Abbey's interior is no less splendid than the exterior, and contains interesting monuments, notably the Despencer and Beauchamp tombs. The Abbey, which was completed in about 1120, was part of a Benedictine monastery until this was threatened with dissolution by King Henry VIII in 1539. The townspeople bought the Abbey—for £453—to save it from demolition, and it became the town's parish church.

The Abbey Cottages Church st. The most unusual buildings in Tewkesbury must surely be the row of mediaeval shops near the Abbey. These 25 cottages were rescued from dereliction and threatened demolition, when it was realised that they are unique in this country. As built, in about 1450, these cottages were made of wattle and daub in a heavy timber framework. The ground floor consisted of trodden earth, the windows had no glass, and there was no chimney in the roof—the smoke from the fire place escaped through a hole under the eaves. The shutters covering the big window facing the street folded down to form a shop counter. One of the houses has been restored to this original state and may be visited; the others have been modified to provide pleasantly discreet modern houses. The restoration of this terrace won a Civic Trust award.

Tewkesbury Museum Barton st. A small museum in a delightfully irregular timber-framed house. Displays of local history, costumes and furniture. Also a large model of the Battle of Tewkesbury. *Open weekdays during the summer. Closed 13.00–14.00.*

Barton Fair takes place in Tewkesbury every 10th October except when that date falls on a Sunday. One of the oldest fairs in the country, this used to be held at the monastery gate.

Tewkesbury Steam Fair and Organ Festival every July, in the meadows by the Avon. This is becoming an important event on steam enthusiasts' calendars, attracting increasing numbers of cherished traction engines, steam rollers and miscellaneous fairground machinery every year.

Battle of Tewkesbury 4th May 1471 Tewkesbury was the last decisive battle in the Wars of the Roses. Margaret, the Queen of the imprisoned Lancastrian monarch Henry VI, had reached England in April, and had immediately hastened to Bristol where she collected arms and a small army. On hearing of her arrival Edward IV, the Yorkist king, organised his army at Windsor and set off westwards to cut off Margaret's advance into Wales, where a large Lancastrian army awaited her. Having failed to make Gloucester, Margaret's tired army reached Tewkesbury on 3rd May, and here Edward caught up with her. The next day the two armies were drawn up on marshy ground to the south of the town, the Lancastrians under the command of Somerset, Wenlock and Devonshire, the Yorkists under Edward, Gloucester and Hastings. The armies were evenly matched. The nature of the ground precluded the effective use of cavalry, and so Somerset attempted a secret right flanking movement to attack the Yorkist rear. His movements were discovered, and his troops scattered in confusion. Somerset then returned to Wenlock, whose troops were holding well, and, accusing him of cowardice in not supporting his movement, he killed him in rage. Leaderless, the Lancastrian army now collapsed into confusion, and fled to Tewkesbury. Prince Edward, Margaret's son, was killed and Somerset was executed the next day. Henry VI was now no longer a danger and so was murdered, leaving the field open for the Yorkist cause.

BOATYARDS & BWB

The following boatyards are all on the Lower Avon in Tewkesbury. There are none on the Severn.

Ⓑ **Beecham Marine** Bredon rd, Tewkesbury, Glos. (293737). Marine engineers: boats built and repaired, inboard and outboard engines repaired. Water, diesel, petrol, gas, chandlery, dustbins, slipway, moorings, winter storage. Extra moorings in new marina off the Avon. Yacht brokers. *Open all year.*

Ⓑ **Avonside Holidays** Gloucester rd, Tewkesbury, Glos. (292284). Boatyard on the Mill Avon. New boat sales and servicing (mostly fibreglass boats). Outboard engine repairs—agents for Penta and Chrysler. Moorings on the Mill Avon, above and below the Abbey Mill. Gas, chandlery and slipway available. *Open all year.*

Ⓑ **William Shakespeare (Boatbuilders)** Avon Boatyard, Tewkesbury, Glos. (292194). Manufacturers of fibreglass motor runabouts, for water ski-ing and racing.

Ⓑ **Tewkesbury Marine Services** St. Mary's lane, Tewkesbury, Glos. (292187). Hire cruisers ('Gay Boats') for canal and river cruising. Diesel, gas, water point. Slipway, moorings on the Mill Avon. Inboard engine repairs. Boat conversions and fitting out. *Open all year.*

BOAT TRIPS

The 'Avon Belle', a former South Coast boat, runs day trips up and down the rivers Avon and Severn, starting on the Mill Avon at Tewkesbury. Public service, and private charter trips (maximum 45 passengers). Enquiries to Mrs Redane, 185 Queens rd, Tewkesbury, Glos. (294088). Small self-drive boats may also be hired from here, by the hour or by the day.

PUBS

Plenty of excellent pubs and hotels in Tewkesbury, though none on the river Avon itself.

Lower Lode on the Severn, ¾ of a mile below Tewkesbury lock. Slipway by the pub, but mooring is very difficult.

CARAVANS & CAMPING

Avonside Holidays Gloucester rd, Tewkesbury, Glos. (292284). Caravan site near the Mill Avon opposite the Abbey. Overnight stops, caravan sales. Hot & cold water, showers, etc. No tents.

FISHING

Tewkesbury—and Upton—hold most of the usual species, the main sport being provided by the chub. There are also some particularly heavy bream and pike around. The Tewkesbury area is noted too for the twaite shad—best place is the weir pool at the back of Tewkesbury lock. This can be grand sport at the right time of year.

TOURIST INFORMATION CENTRES

Tewkesbury The Crescent, Church street, Tewkesbury. (2706).

Tewkesbury Municipal Offices, Oldbury road, Tewkesbury. (2254).

Upton upon Severn

5 miles

The river Severn continues on its predictable, undramatic course northwards, flanked by wooded banks that prevent any views of the countryside. There are few signs of habitation or human activity apart from boats and anglers. The big steel viaduct carrying the M50 motorway provides a rare feature of interest, but the railway that used to bridge the river at Saxon's Lode 2 miles further upstream has vanished. The significant-looking pipes sticking out of the ground on the east bank at this point betray the existence of an old underground oil depot, but it is now disused. A mile further on, things improve as the old church tower at Upton upon Severn comes into view, followed by the graceful modern bridge and interesting waterfront of this very attractive small town. Plenty of boats are moored here – there are temporary public moorings on the west bank, just downstream of the bridge.

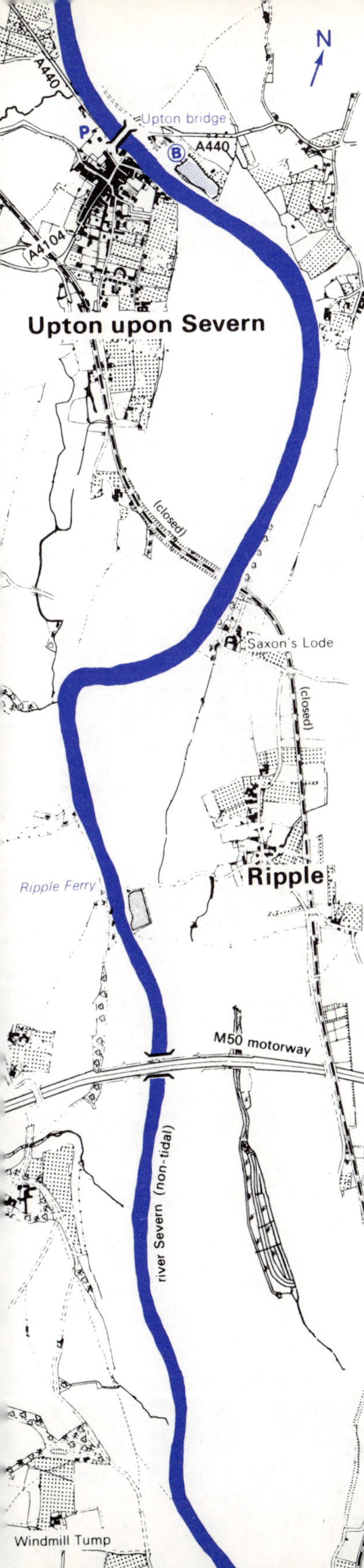

Upton upon Severn
Worcs. Pop 2,000. EC Thur. PO, tel, stores, garage, banks. This delightful town is well provided with fascinating old timbered and early Georgian buildings, and it is doubly welcome for being situated on the river bank. The best area is nearest the river, where various pubs and venerable hotels beckon; nearby is the prominent 13thC tower with its 18thC copper-covered cupola, all that remains of the church which was demolished in 1937. The tower and the former churchyard have recently been restored as a public garden. The 'new' church of SS Peter and Paul was built in 1878, on the edge of the town – a good place for it. There are good shops in Upton, including a delicatessen, and it is the best place along the Severn (apart from Worcester) to forsake a boat for a trip to the famous Malvern Hills, which rise to the west. Great Malvern is under 6 miles to the north-west, and the hills are visible from the Severn.

Ripple
Worcs. PO, tel, stores. Although set back from the river, Ripple is well worth a visit. It is a pretty village, the houses scattered irregularly along the road. The large church is very fine and dates almost entirely from the late 12thC. Only the chancel is late 13thC. The central tower at one time had a spire. Inside are 15thC choir stalls, carved with astrological symbols.

BOATYARDS & BWB

Ⓑ **G. H. Dick (Upton Marine)** Waterside, Upton upon Severn, Worcs. (Upton 2370). Dinghy servicing & chandlery.

PUBS

The only riverside pubs on this section are in Upton.

Plough Inn Upton, near the bridge.
Star Hotel Upton. Old free house near the river. Lunches and dinners daily. Residential. (Upton 2300).
Swan Upton, beside the river.
Ye Olde Anchor Inn Upton, near the church. This pub is dated 1601.
White Lion Hotel High Street, Upton. (Upton 2551).
Railway Inn Ripple.

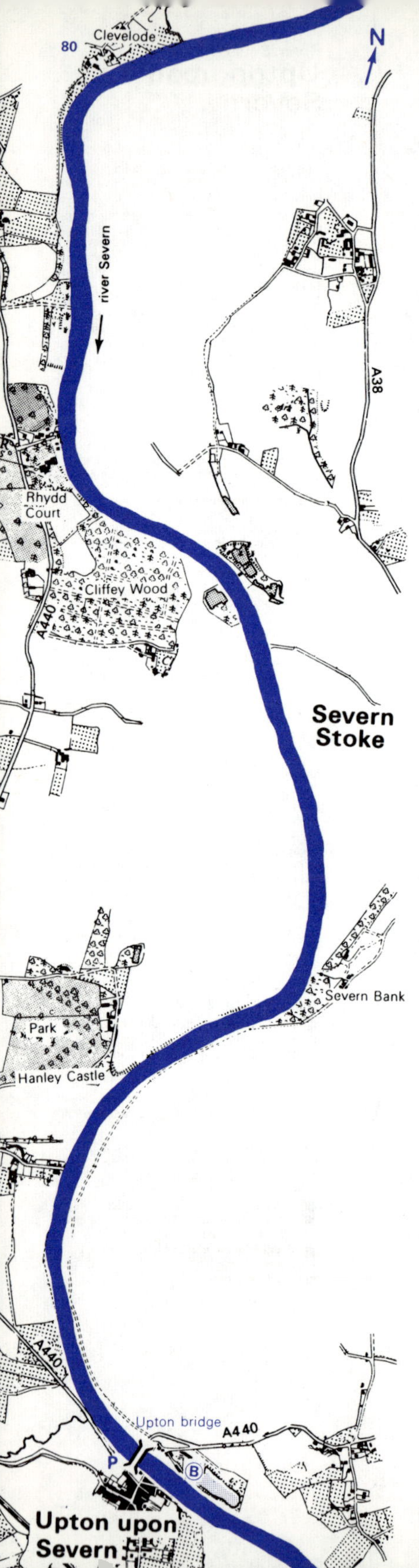

Hanley Castle

5¾ miles

Leaving Upton, the river resumes its high-banked course through the countryside. On the west bank, but hardly visible from a boat, is the village of Hanley Castle. Further up, on the east bank, is a wooded ridge with a curious turreted house projecting from the trees. The village of Severn Stoke is to the east; it is reached by a lane from a jetty on the river. West of here is an enjoyably romantic stretch of river, where tall, steep red cliffs rise sharply from the water to over 100ft. Trees and shrubs struggle to grow from this treacherous slope, and somewhere hidden at the top is Rhydd Court. 'Rhydd' is Welsh for 'ford', but it would seem an unlikely place for a river crossing. The steep hill recedes, allowing a large caravan site to nestle by the river. A scattering of bungalows appears; then the river leaves houses and hills and wanders off north-east. Distantly, to the west, can be seen the grey lumps of the Malvern Hills.

Severn Stoke
Worcs. PO, tel, stores. The village is scattered along the main road, and has no real centre. The best part is near the pretty half-timbered pub with its rose garden. Nearby is the church with its curious 14thC side tower.

Hanley Castle
Worcs. Pop 1,200. PO, tel, stores. The early 13thC castle, built by King John, has now vanished, leaving only its moat as a memorial. But the village that grew up around it still thrives. It is extremely pretty, with a good collection of half-timbered and brick cottages around the little green. Overlooking the green is the church, set in an attractive churchyard. It is a curious building, half 14thC stone, half 17thC brick, with a squat brick tower. Nearby are the 17thC almshouses and grammar school, the whole group a remarkable indication of village life long ago. A lane leads to the village from the river, but mooring is difficult.

PUBS

Rose & Crown Severn Stoke. Off the main road near the church.

Three Kings Hanley Castle. In the village centre. Food, B&B. (Upton upon Severn 2686).

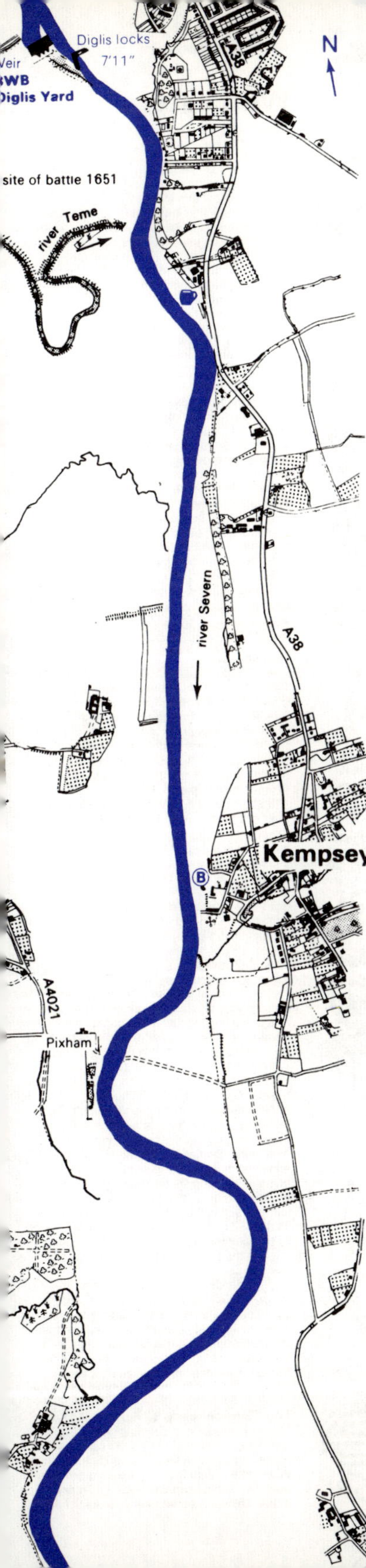

Kempsey

5 miles

The river winds past the hamlet of Pixham, then the bold tower of Kempsey church appears on the east bank, and a line of moored boats betrays a combined boatyard and caravan site. There are reasonable temporary moorings along here for visitors to the village. Upstream, the river straightens out as it makes for Worcester. The Malvern Hills may be glimpsed occasionally, forming the western horizon. Soon the Severn narrows somewhat as a wooded ridge encroaches from the east. The Severn Motor Yacht Club is based here—it is well-named, for the cruisers moored along here are lavish and grand. There is a pub up among the trees near the club. The battle of Worcester was fought in 1651 near where the little river Teme joins the Severn. Above here is the pair of Diglis locks, on the outskirts of Worcester. There is a BWB maintenance yard and a large freight depot above the locks, so dredgers and tugs are often seen.

Battle of Worcester, 3rd September 1651
On 22nd August 1651 the young Charles Stuart (later Charles II), having been proclaimed king by the rebels at Scone, reached Worcester with his Scottish army of 17,000. The Roundhead General Lambert was sent off in pursuit with his northern cavalry, and captured the Severn bridge at Upton upon Severn, cutting off Charles' retreat. Meanwhile another army of 28,000 under Cromwell advanced on Worcester from Nottingham. Charles, realising that he would have to fight at Worcester, organised his defences around the rivers Severn and Teme. After receiving further reinforcements from Banbury, the Roundhead armies advanced across the Severn, using a pontoon made of boats; meanwhile their cavalry crossed by a ford south of Powick bridge, on the Teme. Heavy fighting broke out, and Charles' Scottish infantry, taken by surprise, were soon driven back. Charles tried to redeem the battle by leading a brave charge out of the east gate of Worcester; supported by cavalry this would have succeeded, but by this time the Scottish cavalry had fled. Cromwell held his ground and forced the Royalists back into the town, killing many in the narrow streets. This Roundhead victory ended the Royalist hopes; Charles fled with a few followers, and after the famous Boscobel oak episode he made his way back to France.

Kempsey
Worcs. Pop 1,700. PO, tel, stores. A dull village in which acres of new housing have swamped the old. One or two beautiful thatched cottages have survived to defy the invasion of modernity; but it is the church that should be visited, for the enormous scale of this building is matched by interior grandeur. It was constructed to cater for the Bishop of Worcester and his huge retinue—the Bishop's Palace used to stand just a few yards west of the church. Hence the generous proportions of, especially, the chancel and sanctuary. Note the mediaeval glass in the chancel.

BOATYARDS & BWB

BWB Diglis Maintenance Yard Diglis lock (Worcester 22895). Water.

Ⓑ **Seaborne Yacht Company** Kempsey (295). Court Meadow. Boat builders & hirers; slipway & moorings.

PUBS

Ketch Inn on A38 overlooking the river.

CARAVANS & CAMPING

Seaborne Yacht Co. Court Meadow, Kempsey. (295). Riverside. Overnight stops for touring caravans and motor caravans. Caravans for hire and sale, gas, showers (h&c), shop. *Open Mar-Oct.*

Worcester

4¾ miles

Just above Diglis locks is the terminal basin where the oil tankers used to come to unload before the traffic finished a few years ago. A few hundred yards on are the disused wharves and the two locks that lead into Diglis Basin, the Worcester & Birmingham canal. (*See page 101*). Worcester Cathedral is well in view now; the big square tower commands the town and the riverside. 2 other church towers contribute to the scene, and the unspoilt nature of the west bank makes Worcester's riverside a pretty one. Anglers fish from a path along the east bank, seemingly just below the great west window of the cathedral, while 'fours' and 'eights' appear from rowing clubs. There are two bridges over the river in Worcester–a five-arched stone road bridge and, just north of it, a curious iron railway bridge. The best temporary moorings are between these bridges, on the east bank. North of the railway, the west bank is built up while the east bank is green and tree-lined–a race course is just over the bank. At its northern end is a busy waterworks, contrasting with the bijou houses that adjoin it. North of here the river moves out into pleasantly wooded country; the only trace of civilisation is the glimpse of the occasional farm and a pretty, secluded riverside pub with good moorings. There is a field of hops nearby.

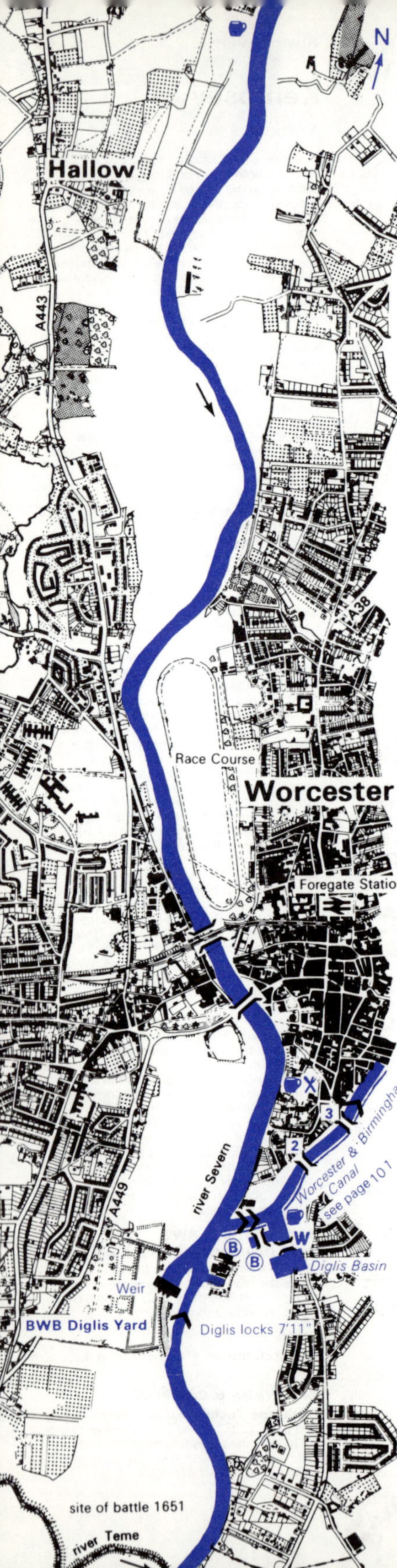

Worcester

Pop 71,000. EC Thur, MD Sat. 2 stations. The 'Faithful City', Worcester has shown loyalty and devotion to the Crown for the last 900 years. Hence the name of Royal Worcester Porcelain, which is still one of the biggest firms in the town. Other industries include glove manufacturing and the making of a certain brown sauce. Worcester has plenty to offer the visitor, although one's enjoyment is lessened by the constant flow of heavy traffic through the town. Foregate street (the main street) suffers worst from the motorists' attention–a pity, because it contains plenty of irregular Georgian buildings with attractive pediments. A railway bridges the street, but does not intrude, for the girders are suitably decorated and trains are infrequent. However the best area is around Friar street, and of course the cathedral.

The Cathedral

An imposing building that dates from 1074 (when Bishop Wulstan started to rebuild the Saxon church), but has work representative of the five subsequent centuries. There is a wealth of stained glass and monuments to see–including the tomb of King John, which lies in the chancel. Carved out of Purbeck marble in 1216, this is the oldest royal effigy in England. When he was dying at Newark, King John demanded to be buried at Worcester Cathedral between 2 saints; but the saints have gone now. The best way into the Cathedral is from the Close with its immaculate lawns and houses, passing through the cloisters where one may inspect five of the Cathedral's old bells. (Two of these were cast in 1374). The gardens at the west end of the building look out over the Severn and over to the Malvern Hills–a particularly fine sight at sunset.

The Three Choirs Festival is held annually in rotation at the cathedrals of Worcester, Gloucester and Hereford, during the last week in August. This famous festival has inspired some fine music, one notable composer being Vaughan Williams. For further information about the festival contact the Town Clerk in any of the three cities.

The Commandery by Kings Head lock, on the Worcester & Birmingham Canal. (Access for the agile from the canal). This was founded as a small hospital by Bishop Wulstan in 1085, but the present timbered building dates from the reign of Henry VII in the 15thC. It served as Charles Stuart's headquarters before the Battle of Worcester in 1651. The glory of the building is the superb galleried hall with its ancient windows and the Elizabethan staircase. *Open daily 10.00–16.30.*

Tudor House Folk Museum Friar street (Worcester 22154). A new museum of local antiquities, furniture and porcelain housed in traditional Elizabethan buildings. (The wattle and daub that makes up the walls can be clearly seen in places). Amongst other exhibits are a modern copy of a traditional coracle, a tiny fishing craft used for thousands of years on the river Severn, and a painting of the Waterman's church in Worcester—a chapel on a floating barge, last used in the 1870's, when it was taken ashore and set up on dry land. *Open 10.30–17.00 weekdays; closed Thur and Sun.*

The Dyson Perrins Museum of Worcester Porcelain The Royal Porcelain works, Severn st. (23221). Here, where it should be, is the most comprehensive collection of Worcester porcelain in the world, from 1751 to the present day. *(Open Mon-Fri 10.00–13.00, 14.00–17.00, also on Sat Apr-Oct. Tours of the porcelain works can be arranged Mon-Fri at 10.00 and 14.00).* It is advisable to book for tours.

The Guildhall High st. Built in 1721–3 by a local architect, Thomas White, this building has a splendidly elaborate façade with statues of Charles I and Charles II on either side of the doorway and of Queen Anne on the pediment. It contains a very fine assembly room.

City Museum and Art Gallery Foregate st. (22154). Collections of folk life material and natural history illustrating man and his environment in the Severn Valley. In the Art Gallery are a permanent collection and loan exhibitions. Also museum of the Worcestershire Regiment. *Open Mon-Sat 9.30–17.00.*

The Greyfriars Friar st (NT property). Dating from 1480, this was once part of a Franciscan Priory and is one of the finest half timbered houses in the country. Charles II escaped from this house after the Battle of Worcester on 3rd September 1651. It is an antique shop now.

BOATYARDS

Several in Diglis Basin (*See page 101*), but none on the Severn itself.

BOAT TRIPS

Worcester Steamer Company Diglis, Worcester (27543). Large motor trip boats operating along river Severn, for party bookings only (up to 200 passengers carried). Buffet & music can be arranged.

PUBS

Camp House Inn Grimley. An isolated riverside pub below Bevere lock. Snacks at the bar—full meals only with 48 hrs notice. Good moorings. (Worcester 640288). The drinking water here is pumped up electrically from a well.

Crown Hallow. Restaurant (on the A443).

Severn View hotel Worcester. Near the river by the railway bridge.

Old Rectifying House Worcester, by the road bridge. B&B.

Diglis Hotel Worcester, on the river by the Cathedral. Restaurant. (Worcester 27978).

TOURIST INFORMATION CENTRE

Worcester The Guildhall, Worcester. (23471).

Worcester Cathedral from the river Severn.

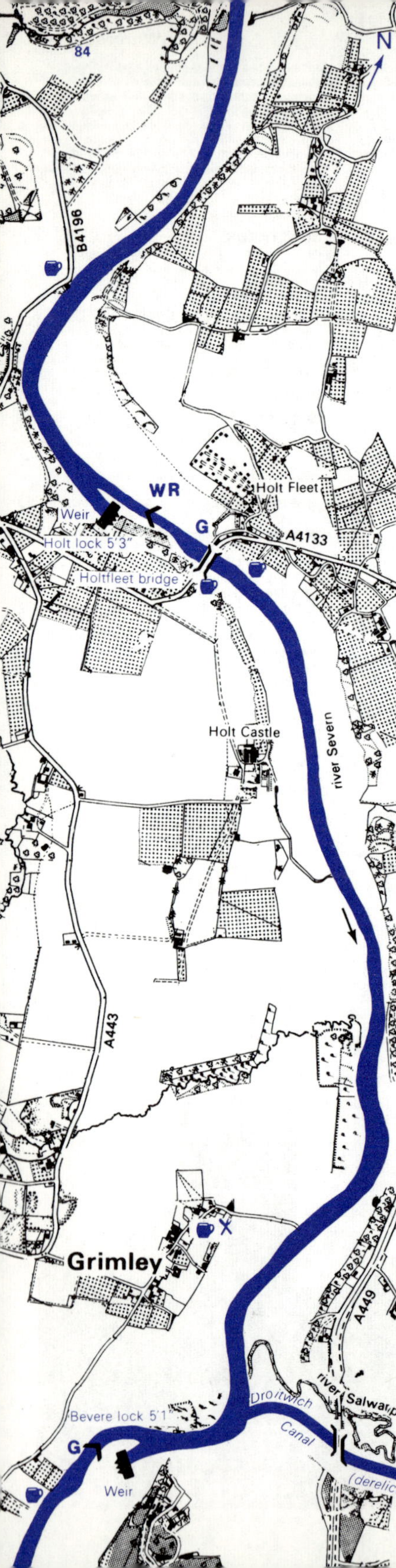

Holt Castle

$5\frac{1}{2}$ miles

Just upstream of the Camp House Inn is Bevere lock, which is certainly one of the prettiest on the Severn. There is a delightful rose garden, tended by the lock keeper and his wife. The island adjoining the lock is connected to the land on the far side by a graceful iron footbridge. Continuing upstream of the lock, the river is approached by wooded hills on its east bank, while on the other side the plain green fields continue, edged by high banks and trees. At one point the little river Salwarpe and the disused Droitwich Canal enter together from the east. The village of Grimley is at the end of a lane leading up from the river, but it is difficult to distinguish this track, and moorings are not available. The river continues north-westward now, until Holt Castle is reached, a curious composite building overlooking the river. Beside it is the discreet tower of a small church. Further on is the delicate iron span of Holt Fleet Bridge, and beyond it a steep wooded hillside, with pubs and caravans nearby. Upstream of the bridge is Holt lock—water and dustbins are available for boats here. Above the lock, the steep wooded hills continue, rising straight up from the river bank. It is a pleasant scene, and there is a riverside pub nearby. The next few miles form an attractive reach, with steep wooded hills rising from the river bank on one side, then on the other. There is a riverside pub along here, with a small settlement of inoffensive-looking chalets.

Holt
Worcs. PO, tel, stores. Holt is a scattering of assorted settlements, around the river. The elegant narrow bridge here was built by Telford in 1828. Holt Fleet exploits the river in a most unattractive fashion, being composed of a large sprawling caravan site, whose television aerials and wires swamp the riverside. There are two large pubs. Holt Castle is up on the hill, overlooking the river. The 'castle' is a 14thC tower, for the rest of the building is a 19thC battlemented mansion. Nearby, set among the fruit fields, is the church, a fine late Norman building with interesting carving around the doorways, and rich in interior ornamentation.

Grimley
Worcs. PO, tel, stores. A small farming village close to, but hidden from, the river. The well placed church has some Norman work, but has been heavily restored, it has a curious outside staircase by the door. Access from the river is not easy, but there is a pub.

The Droitwich Canal
This attractive rural waterway, which leaves the Severn $\frac{1}{2}$ mile NE of Bevere lock, used to go to Droitwich—$5\frac{3}{4}$ miles and 8 locks away. It was then joined by the Droitwich Junction Canal, whose 7 locks led it a further $1\frac{1}{2}$ miles to terminate in a junction with the Worcester & Birmingham Canal at Hanbury Wharf (*see page 103*). Both the Droitwich canals have been derelict for most of this century, but they are now reviving. The local council, Droitwich Town Development, is gradually implementing a scheme to open up the Droitwich Canal from the town to the river Severn, and it is hoped later to extend the restoration to the Junction Canal. When and if this happens, it will restore a 22-mile ring of cruising waterways.

PUBS

Lenchford Holt. Riverside, upstream of Holt lock.

Holt Fleet Hotel Holt Fleet, by the bridge.

Wharf Hotel Holt Fleet, north bank by caravan site.

X **Waggon Wheel** Grimley. Free house. Food.

CARAVANS & CAMPING

Holt Fleet Farm Holt Fleet. (Ombersley 512). Overnight stops for touring caravans and tents. *Open Apr-Oct.*

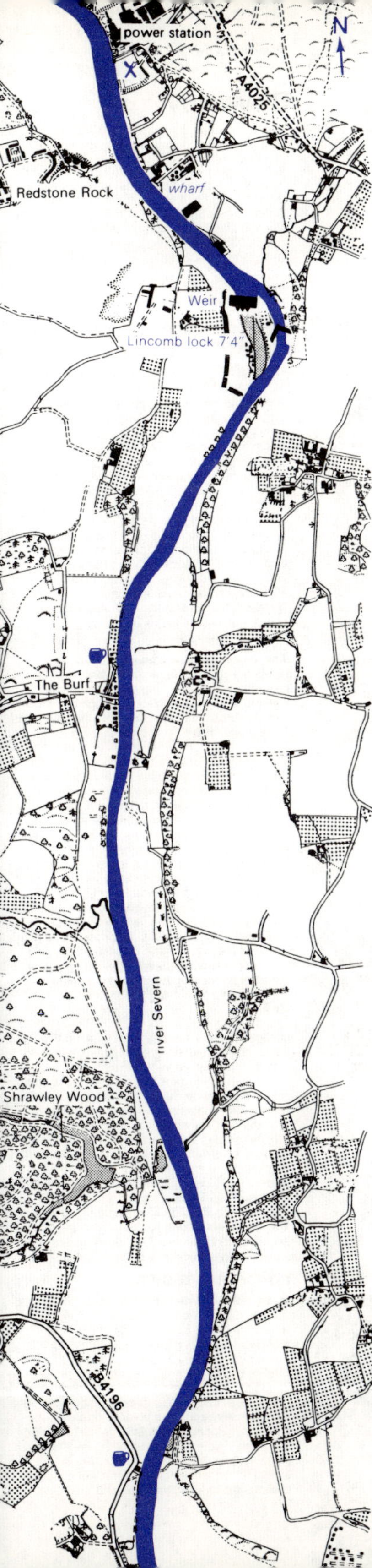

The Burf

4¾ miles

The reach from Lenchford to Stourport is one of the most pleasant on the Severn. Unlike much of its journey further downstream, the river runs here through a well-defined valley, with steep wooded hills never far away from either bank. The hills on the west bank are the more impressive and the more thickly wooded, although the old church at Shrawley can sometimes be seen peering over the woods. Roads keep their distance, but at the site of Hampstall Ferry (The Burf) there is a small village and a riverside cider house, with good moorings. From here very steep hills encroach on the east bank of the river, almost hanging over it at Lincomb lock. This pretty lock is now the northernmost on the river, for signs of Stourport soon come into view. First there are the abandoned oil wharves, whose rusty pipes and terminals are a sad reminder of the former traffic. On the opposite side of the river, at the foot of a cliff, is the Redstone Rock—an unexpected outcrop of crumbling red sand-stone. There was once an hermitage in caves here. On towards Stourport, there is a modest restaurant on the east bank and, beyond it, Stourport Power Station. The river Stour flows in here.

The Burf
An isolated riverside settlement, comprised mainly of new housing for retired persons.

PUBS

Fountain Grill & Snack Bar, on east bank of river below Stourport Power Station. Licensed restaurant with cruisers for hire. Moorings.

Hempstall Cider House The Burf, overlooking the river. Good moorings. Food (and beers).

Bewdley

5 miles

At Stourport Power Station the little river Salwarpe flows in from the east, and just to the north the Staffordshire & Worcestershire Canal drops down into the Severn from the unseen basins. There are two sets of locks—narrow canal boats should use the upstream set. Just above these locks is Stourport Bridge, a heavy iron structure built in 1870. Boats are moored on either side of this bridge and a boatyard is situated here. *(For notes on Stourport and the Staffordshire and Worcestershire Canal see page 89).*
The River Severn is not officially navigable for more than a couple of hundred yards above Stourport Bridge, at which point BWB's jurisdiction as navigation authority ends. However in suitable conditions small boats not drawing more than 1ft. 9in. can penetrate upstream to within a mile of Bewdley Bridge. A shoal across the river impedes further progress, and boatmen must tie up or wade ashore when they reach the shoal. There is a road on either side of the river into Bewdley on either side of the river (the B4194 on the west bank being the most direct). Alternatively, one can avoid all this by leaving the boat in Stourport and taking a bus to Bewdley. (Midland Red runs services 312 and 314 on weekdays along the east side of the river). An excursion to Bewdley is well worth the effort.

The river Severn above Bewdley
In the 19thC the Severn was fully navigable for a long way past Stourport: it used to be a vital trade artery right up into Wales, through Bridgnorth, Shrewsbury and Newtown. It used to connect with the Montgomery Canal at Newtown, the Shrewsbury Canal at Shrewsbury, and the Shropshire Canal at Hay. It is unfortunate that this upper section is unnavigable, for the river is much prettier than further south: it runs along a narrow valley, hemmed in by steep and wooded hills. However there is a public right of way along one or both banks up to Bridgnorth and beyond, and this can form an interesting walk. There is a Midland Red bus service from Upper Arley (number 296) and Bridgnorth (number 297) to Kidderminster, from where buses serve Bewdley and Stourport. Another attraction north of Bewdley is the Severn Valley Railway, a private railway which runs a summer service of steam trains between Bridgnorth and Hampton Loade.

Bewdley
Worcs. Pop 6,000. EC Wed. PO, tel, stores, garage, bank. Bewdley is a magnificent small 18thC riverside town, still remarkably intact. It is blessed with a fine river frontage and elegant bridge that make the most of the wide Severn, and a handsome main street that leads away from the river to terminate at the church. The scale of the whole town is very pleasing, a comfortable mixture of old timber-framed buildings and plainer, more elegant 17thC and 18thC structures. Most of the town is on the west bank, and so Telford's three arch stone bridge, built in 1798 to replace an earlier mediaeval structure, forms a fitting entrance to Bewdley. Old warehouses, hotels, and grand houses flank the quays.

BOATYARDS & BWB

Ⓑ **Stirchley Marine** Stourport, just above the bridge. Office and enquiries: 1240 Pershore Road, Stirchley, Birmingham. (021 458 3946). Hire cruisers; permanent moorings. Dayboats (motor runabouts) for hire by the hour.

Ⓑ **Head's Boatyard** Riverside, Stourport on Severn. (2044). Two establishments, one on river Severn by the bridge, one in Stourport Basin, above the locks. (*See page 90*). The yard on the river offers: water, gas, slipway, moorings, winter storage, boat sales (fibreglass and wooden river & canal cruisers up to 25ft. built here).

Staffordshire & Worcestershire

Maximum dimensions

Length: 72′
Beam: 7′
Headroom: 6′ 6″

Mileage

STOURPORT to:
Kidderminster lock: 4½
Wolverley lock: 6
STOURTON JUNCTION: 12¼
Swindon: 16¾
Bratch locks: 19
ALDERSLEY JUNCTION: 25
AUTHERLEY JUNCTION: 25½
GREAT HAYWOOD JUNCTION: 46

Total 31 locks

Construction of this navigation was begun immediately after that of the Trent & Mersey, to effect the joining of the rivers Trent, Mersey and Severn. After this, only the line down to the Thames was necessary to complete the skeleton outline of England's narrow canal network.

Engineered by James Brindley, the Staffs & Worcs was opened throughout in 1772, at a cost of rather over £100,000. It stretched 46 miles from Great Haywood on the Trent & Mersey to the river Severn, which it joined at what became the bustling canal town of Stourpourt. The canal was an immediate success. It was well placed to bring goods from the Potteries down to Gloucester, Bristol and the West Country; while the Birmingham Canal, which joined it halfway along at Aldersley Junction, fed manufactured goods northwards from the Black Country to the Potteries via Great Haywood. Stourport has always been the focal point of the canal, for the town owed its birth and rapid growth during the late 18th century to the advent of the canal. It was here that the cargoes were transferred from narrowboats into Severn Trows for shipment down the estuary to Bristol and the south west.

The Staffordshire & Worcestershire canal soon found itself facing strong competition. In 1815 the Worcester & Birmingham canal opened, offering a more direct but heavily locked canal link between Birmingham and the Severn. The Staffs & Worcs answered this threat by gradually extending the opening times of the locks, until by 1830 they were open 24 hours a day. When the Birmingham & Liverpool Junction canal was opened from Autherley to Nantwich in 1835, traffic bound for Merseyside from Birmingham naturally began to use this more direct, modern canal, and the Staffs & Worcs lost a great deal of traffic over its length from Autherley to Great Haywood. Most of the traffic now passed along only the ½-mile stretch of the Staffs & Worcs canal between Autherley and Aldersley Junctions. This was, however, enough for the company, who levied absurdly high tolls for this tiny length. The B & LJ company therefore co-operated with the Birmingham Canal company in 1836 to promote in Parliament a Bill for the 'Tettenhall & Autherley Canal and Aqueduct'. This remarkable project was to be a canal 'flyover', going from the Birmingham Canal right over the profiteering Staff & Worcs and locking down into the Birmingham & Liverpool Junction canal. In the face of this serious threat to bypass its canal altogether, the Staffs & Worcs company gave way and reduced its tolls to a level acceptable to the other 2 companies. In later years the device was used twice more to force concessions out of the Staffs & Worcs.

In spite of this setback, the Staffs & Worcs maintained a good profit, and high dividends were paid throughout the rest of the 19thC. When the new railway companies appeared in the west midlands, the canal company would have nothing to do with them; but from the 1860s onwards, railway competition began to bite, and the company's profits began to slip. Several modernisation schemes came to nothing, and the canal's trade declined. Like the other narrow canals, the Staffordshire & Worcestershire faded into obscurity as a significant transport route by the middle of this century, although the old canal company proudly retained total independence until it was nationalised in 1947. Now the canal is used almost exclusively by pleasure craft—and it is certainly a most delightful canal for cruising (or walking), especially in the southern reaches in the sandstone area.

Wild flowers

The Staffordshire & Worcestershire Canal provides a very attractive long-distance walk, and one of particular interest to nature lovers. Most of the towpath is in excellent condition and provides a lovely walk, especially along the sandstone cuttings near Wolverley. Here the walker is often preceded for miles by low-flying herons.

Probably the most conspicuous plant along this canal is the Himalayan balsam. It was introduced in about 1840 and the valley of the Smestow brook was one of the first places to be colonised. Although this is an annual plant, it grows up to ten feet high and has beautiful flowers ranging from pinkish-purple to white. The seeds are dispersed by an explosive mechanism—try to gather some ripe seeds and you will realise this! Although primroses are not found, in spring the towpath is a blaze of colour with the yellow lesser celandine and the white cow parsley with its fern-like leaves. Bluebells are uncommon but if you look across west from below Swindon lock you will see the woods covered with a mass of blue. Generally the earliest flower to appear is coltsfoot—the purplish stems with many scales bearing dandelion-like flowers and appearing before the leaves. This is an old-fashioned remedy against sore throats and many countryfolk consider 'coltsfoot rock' a pleasant and efficacious cure. Another early flower is stitchwort, of rather grass-like appearance and with beautiful white flowers.

The canal is accompanied by hawthorn hedges in which grow wild roses and sometimes the wild cherry. Often the bushes are covered with the climbing white bryony and its red berries in the autumn. The large convolvulus has lovely flowers, generally white but occasionally various shades of pink. Sometimes everything is smothered with the wild clematis or traveller's joy, the seeds of which are covered with white plumes; these give the plant the alternative name of 'old man's beard'. Nestling at the bottom of the hedge grows 'lords and ladies'; this sends up arrow-shaped leaves in the spring, then a pale green hood and finally a short spike of orange-red berries. You will often find scrambling up the hedge the woody nightshade, sometimes mistakenly called deadly nightshade—a plant which does not grow here. The flowers are an attractive mixture of purple lobes and yellow anthers. This plant is slightly poisonous.

Everyone who walks the canal will be fascinated by the narrow locks, and it is surprising how many plants grow on the lock walls. Perhaps the most beautiful is the small skullcap with its pairs of bright-blue flowers. A beautiful flower that is spreading quickly is rose-bay, with its tapering spikes of pinkish-purple flowers and with long silky hairs attached to the seeds. In wet places grows an allied plant, 'codlins and cream', with larger flowers. Nettles, of course, unfortunately abound but they do support the caterpillars of the lovely peacock butterfly—which, together with the small tortoiseshell and orange-tip, is the butterfly most commonly found on this canal. An interesting flower in early spring is butterbur, which sends up short heads of lilac-pink flowers. Later come the very large leaves, on stout stalks; these give the plant the common name of 'wild rhubarb'. In the waterway itself the commonest plant is water plantain with branched spikes of lilac-white flowers with three petals. Close to the water grows common valerian—very different from the red-flowered variety that flourishes on walls. This has small, very pale pink flowers. Near this you will often find bur-marigold with nodding heads of a deep yellow. Another water loving plant is water figwort, a tall plant with square stems and small reddish-brown flowers. Of the scented flowers perhaps the most attractive is meadowsweet with clusters of tiny cream-coloured flowers.

In some places grows the hemlock, a tall plant with spotted stems, beautiful and finely cut leaves and clusters of small white flowers. It was the poison from this plant that the Athenians used to execute the philosopher Socrates in 399 B.C. A plant that flourishes almost anywhere is mugwort (a plant once used to give a pleasant flavour to mugs of ale). This is tall and branched with small oval yellow flowers. Allied to this is wormwood, with greyer leaves, more aromatic and slightly larger flowers. From this family is made the French liqueur absinthe.

Some plants will be found in flower throughout the year whenever conditions are suitable. These include the white dead-nettle, groundsel (so well known to gardeners!) and oxford ragwort, a bushy perennial with glossy, lobed leaves and bright yellow flowers. This originally grew on old walls at Oxford but started to spread last century. It was probably helped by the Great Western Railway—the ballast formed an ideal site for the plant and the rushing trains helped to spread the seeds. It is still found growing on some college walls. 'Common' ragwort is taller and stiffer; it flowers only during the late summer. If you are interested in uncommon flowers then in the southern section of the canal you may find shepherd's rod, with small round heads of white flowers and with purple anthers.

Growing almost everywhere on the towpath are members of the geranium family, their flowers generally pink and the fruit ending in a long pointed beak, giving rise to the common name of cranesbill. Varieties include herb robert with fern-like leaves, dovesfoot with round leaves and small-flowered cranesbill with the leaves divided. A tall-growing plant is mallow, with crinkly leaves and pinkish-purple petals with darker veins. The fruits—the so-called cheeses—are arranged in a ring. Another common flower is St. John's wort, about two feet high with golden yellow flowers. The leaves have minute perforations. Formerly the plant was looked upon as holy and a protection against the Devil. This so enraged the Devil that he stabbed the leaves through and through with a needle!

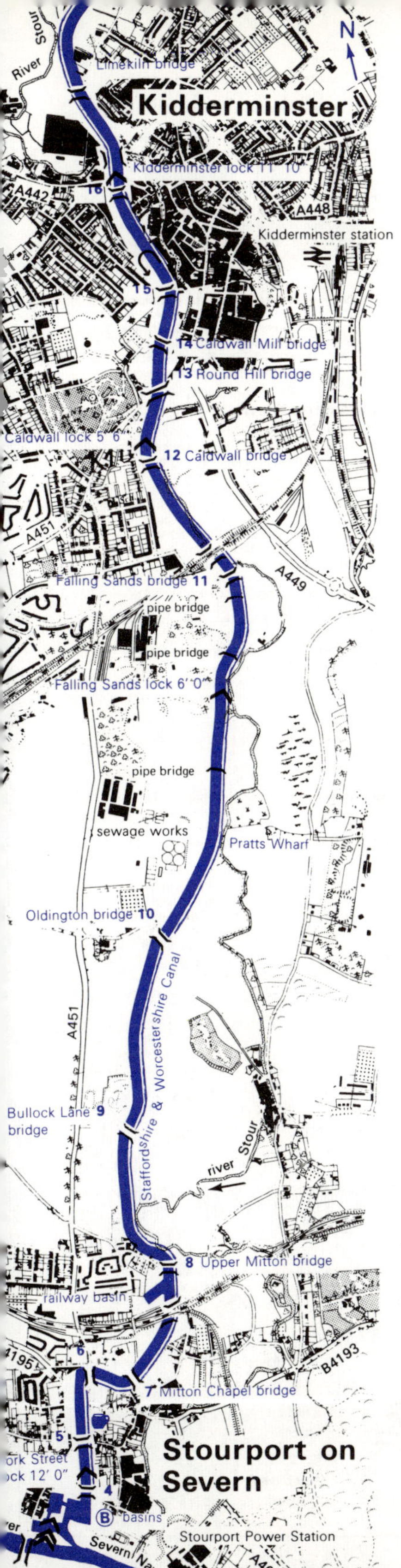

Staffordshire & Worcestershire

Stourport and Kidderminster

$4\frac{1}{2}$ miles

Few of the boatmen who quit the river Severn to navigate the Staffs & Worcs canal will regret their decision. The lazy, monotonous and sweeping course of the river is replaced by the fascination and ever-changing scenery of a narrow canal, threading its intimate way through pleasantly irregular countryside. This canal is, without doubt, one of the prettiest waterways in England.

First one must negotiate the locks and basins at Stourport—a pleasant task, for the combination of all kinds of engineering features, fine buildings and pleasure boats makes this a lively, colourful scene. The famous clock tower looks out over all. There are two sets of locks here, narrow and broad (the former have only recently been restored into good working order). *All the locks here are kept padlocked: the lock keeper, whose cottage is at the top wide lock, is on duty every day from 8.00 to 20.00, except from 12.00 to 13.00 and 17.00 to 18.00. (Telephone Stourport 2838). For an enlarged map of the Stourport basin area see overpage.* To reach the Staffs & Worcs canal itself, boats should proceed to the eastern corner of the upper basins, pass under the bridge and climb the deep lock at York Street. There is a small studio at this lock selling painted canal wares. Above the lock are good temporary moorings, opposite a boatyard which has carefully preserved and utilised the old canal buildings. Although it is still in the middle of Stourport, the canal has already acquired a secluded, unspoilt character, flanked by discreet, gently decaying houses and walls and accompanied by a rural towpath. The navigation seems to creep through the town, twisting along past the big disused Milton railway basin at a railway bridge and emerging quickly into the country. It follows the west side of a little valley, the steep slopes rising sharply from the water. The river Stour approaches and at the derelict Pratt's Wharf the towpath rises over an almost unrecognisably overgrown lock that once joined the canal to the Stour. This river used to be navigable from here for $1\frac{1}{4}$ miles down to the Wilden Ironworks. The sweet smell from an adjacent sewage works spurs the traveller on towards Kidderminster and suddenly the canal's surroundings change. For the hillside on the west bank becomes a dramatic cliff of crumbling red rock rising sheer from the canal. This is the southern end of a geological feature that stretches almost to Wombourn, 15 miles away. Falling Sands and Caldwall locks both enjov delightful situations at the foot of the sandstone, and both have split iron bridges of a type usually associated with the Stratford on Avon canal. Kidderminster is now reached, and the canal's course through the centre of it is truly private, passing along a corridor of high walls, factories and warehouses that clearly date from the arrival of the canal in Kidderminster. As the buildings mount up, the navigation narrows until it escapes by diving into a short tunnel, at the end of which is a deep lock. One surfaces to find a very different scene: this is open townscape, with traffic all around, shopping streets close by and a church just ahead, on a rise. Nearby is a statue of Richard Baxter, the 17th century thinker who 'advocated unity and comprehension' in religion. Just above Kidderminster lock, the river Stour appears from nowhere: the canal crosses it on an aqueduct. Continuing northwards, the canal curves away to leave the town centre behind, passing now more modern and less picturesque industrial works.

Kidderminster

Worcs. Pop 47,000. EC Wed. MD Thur, Sat. Kidderminster exists above all for carpet weaving. The industry was first introduced here in 1735, when the town was already a prosperous cloth manufacturing centre, and today there are 18 factories in Kidderminster involved in the production of carpets. Born here in 1793 was Rowland Hill, who founded the Penny Post in 1840: his statue in front of the head post office commemorates his 'creative mind and patient energy'. From the canal, one sees little but the older industrial side of Kidderminster, which is not without interest. The best place for access to the town is from Kidderminster lock; the public baths are conveniently beside bridge 15. There is not much to see in the town, which has few buildings of real interest. The centre itself is better now than it was, enjoying the benefit of quiet streets uncluttered by vehicles; but this has been achieved at the expense of widespread demolition to clear the way for a modern dual carriageway—an inner ring road. This road unfortunately passes the door of the impressive, dark church of St. Mary, cutting it off completely from Church Street, in which Kidderminster's few Georgian houses are situated.

Kidderminster Museum & Art Gallery Market st (62832/3). Collection of archaeological finds and exhibits of local interest. Small permanent art collection, including some Brangwyn etchings. (There is much more to see at the Worcestershire County Museum in Hartlebury Castle. *See below*). *Open Mon-Sat 9.30–18.00.*

Stourport-on-Severn
Worcs. Pop 15,000. EC Wed. PO, tel, stores, garage, bank. When the engineer James Brindley surveyed the line for the Staffordshire & Worcestershire Canal in the early 1760's, he intended to meet the river Severn at Bewdley, which was already an established river port. But the residents there objected to his plans, and so he chose instead the hamlet of Lower Mitton, 4 miles downstream, where the little river Stour flowed into the Severn. Basins and locks were built for the boats, warehouses for the cargoes and cottages for the workmen. The canal company even built in 1788 the great Tontine hotel beside the locks. The hamlet soon earned the name of Stourport, becoming a busy and wealthy town. The two basins were expanded to five (one has since been filled in) and the locks were duplicated. Nowadays, plenty remains of Stourport's former glory, for the basins are always full of boats (there is a boat club and a boatyard). The delightful clock tower still functions, a canal maintenance yard carries on in the old workshops by the locks, and the Tontine hotel still has a licence. Mart Lane (on the north-east side of the basins) is worth a look—the original 18thC terrace of workmen's cottages still stands. Numbers 2, 3 and 4 are listed as ancient monuments. In contrast with the basin area, the town of Stourport is not interesting, and although it was built on account of the canal, the town has no relationship at all with the basins now. It seems to have grown up away from the canal.

Hartlebury Castle *2 miles east of Stourport, on the B4193.* The home of the Bishops of Worcester, this castle was built in the 15thC, virtually destroyed in 1646 and rebuilt in the 18thC. *Open Sun and B. Hols 14.00–18.00.*

The Worcestershire County Museum is housed in part of the Castle. It shows the story of the archaeology, geology, crafts and industries of the county. *Open weekdays (except Fri) 10.00–18.00, Sat & Sun 14.00–18.00. Closed Dec and Jan.* (Hartlebury 416).

BOATYARDS & BWB

Ⓑ **Canal Pleasure Craft** Stourport-on-Severn, Worcs. (2970). Hire cruisers. Diesel, petrol. Boat building and sales. Minor boat repairs. Closed Sun. Situated on the canal above York Street lock.

Ⓑ **Head's Boatyard** Stourport-on-Severn, Worcs. (2044). Comprehensive boating service in Stourport Basin and on the Severn by Stourport bridge. Water, gas. Slipway up to 45ft, permanent moorings and winter storage. Canal & river cruisers and dinghies built, in wood or glass fibre. Boats sold and repaired; inboard and outboard engine repairs. Big chandlers shop in Mart Lane, on east side of basins.

BWB Stourport Yard Stourport Basin (2838). Water, refuse and sewage disposal nearby.

PUBS

None on the canal in Kidderminster.

Bird in Hand Stourport. Canalside, south of the railway bridge.

Black Star Stourport. Canalside, by bridge 5.

Bell hotel Stourport, opposite York Street lock.

White Lion near the Bell.

Tontine hotel Stourport Basin.

BOAT TRIPS

Heads Boatyard operate 3 river launches from Stourport. 40 minute public trips at weekends and bank holidays, available for private charter at any other time, day or night. Maximum no. of passengers 199 on each boat. All enquiries to Heads Boatyard (Stourport 2044).

STOURPORT BASINS.
Enlargement from map page 89.

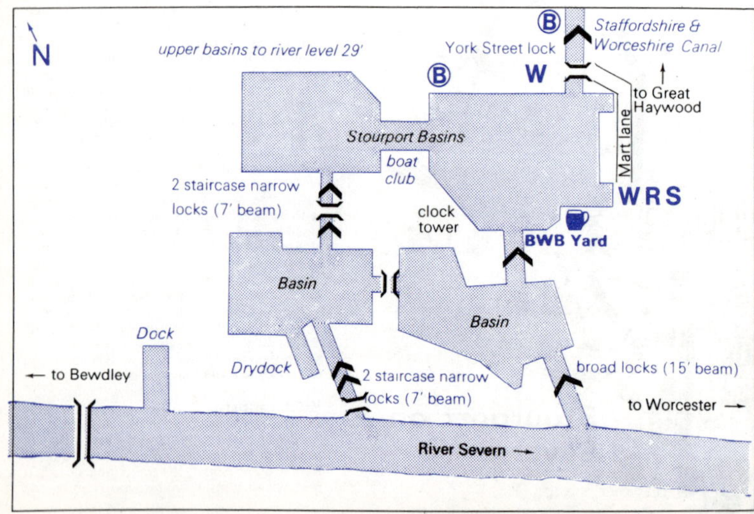

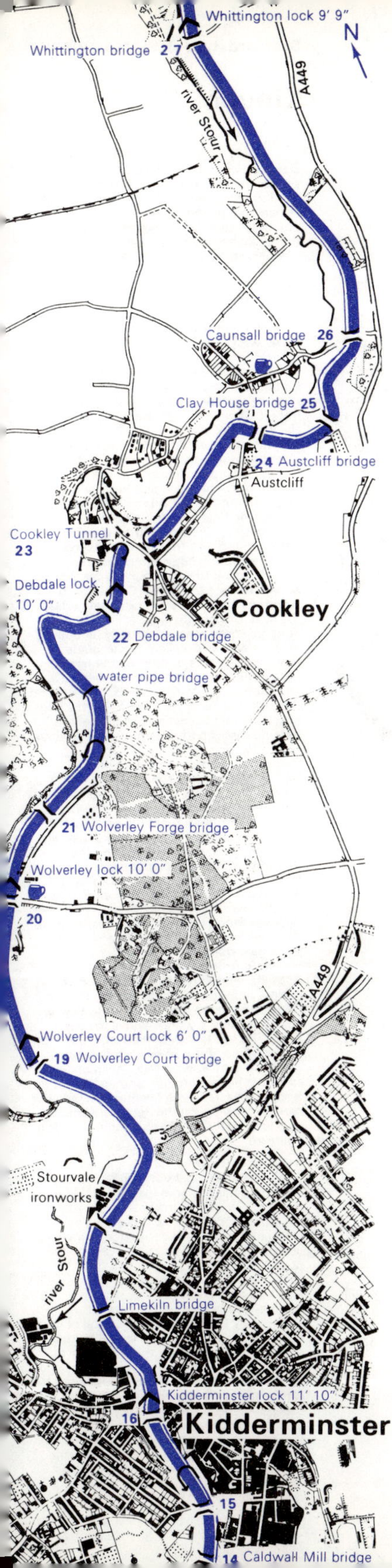

Wolverley

5 miles

This is another delightful stretch of waterway. Leaving Kidderminster, the navigation moves into an area of quiet watermeadows created by the little river Stour, which is now on the west side. Past an isolated lock, the village of Wolverley on the other side of the valley is given away by its unusual Italianate church standing on a large outcrop of rock. The approach to the deep Wolverley lock is lined by trees. There is a pub beside the lock, and good moorings above and below it. Beyond here, the course of the canal becomes really tortuous and narrow as it proceeds up the narrow, thickly wooded valley, forced into endless diversions by the steep cliffs of friable red sandstone. Vegetation of all kinds clings to these cliffs, giving the impression of jungle foliage. At one point the navigation opens out, becoming momentarily like a normal canal; but soon the rocks and trees encroach again, returning the waterway to its previous constricted width. An impressive promontory of rock compels the canal to double back on itself in a great horse-shoe sweep that takes it round to the pretty Debdale lock. A doorway reveals a cavern cut into the solid rock here: this was used as an overnight stable for towing horses. A little further on is an old branch to a (still functional) steel works, and past it is a short tunnel, with the small village of Cookley on top of it. Beyond the tunnel, the steep rocks along the right bank culminate in a remarkable geological feature where, unnervingly, the Austcliff Rock actually leans out, trees, bushes and all, right over the canal. Then the rocks recede for a short distance. Thick woods here keep the nearby A449 road at bay. Across the river Stour, $\frac{1}{4}$ of a mile west of bridge 26, is the small settlement of Caunsall. There are farms here, and two pubs, but little else.

Cookley

Worcs. PO, tel, stores, garage, fish & chips. The village is set well above the canal, which passes underneath it in a tunnel. Although it has an attractive situation, Cookley is not a particularly pretty village, and there is little to visit. The big steel works near the canal produces, among other things, pressed steel wheels for motor vehicles: the biggest bulldozer wheel in the world was made here. Down in the valley, near the river Stour, there are the older, more attractive cottages of the village. And clearly visible are the entrances to caves in the cliff face. To reach Cookley from the canal, it is best to moor west of the tunnel, then walk up the path to the village.

Wolverley

Worcs. PO, tel, stores. 400yds NW of bridge 20. A very unusual and pretty village on the west side of the Stour valley. The church is predominant, a dark red structure built in 1772 in a precise Italianate style. It stands on a sandstone rock so steep that the building has to be approached by a zig-zag path cut through the constantly eroded sandstone. Most of the village is clustered just to the north of the church, near the little-used but dignified stone buildings of the old grammar school. The school was endowed in 1629, but most of the buildings date from 1820. Around it is the bulk of this small village, where gardens make the most of the brook that flows through. There is an attractive pub in the centre, and another, with spacious gardens, up the hill. Wolverley is a village certainly worth visiting—it is easily accessible from Wolverley lock.

PUBS

- **Rock** Caunsall, $\frac{1}{4}$ mile W of bridge 26.
- **Queens Head** Caunsall, near the Rock.
- **Bulls Head** Cookley. Up the hill from the tunnel, near the shops.
- **Lock** Canalside, at Wolverley lock. Terrace fronting the canal. Food usually available.
- **Queens Head** Wolverley, near the old school. Terrace.

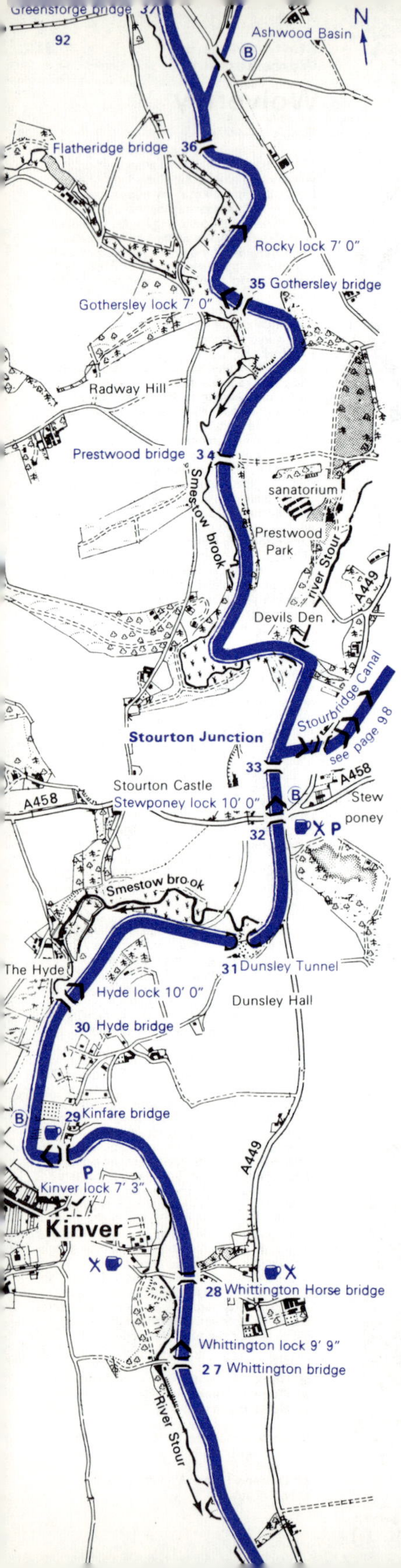

Kinver

5½ miles

Between bridges 26 and 27 the canal passes from Worcestershire into Staffordshire, but the surroundings of this remarkable waterway do not change: it continues through secluded woodlands, the rocky hillside on the east bank steepening as the valley narrows again. The nearby main road remains unnoticed, while the canal reaches Whittington lock, a pretty lock cottage beside it. The bridge at the lock tail is typical of this navigation, its parapet curving fluently round and down to the lower water level. A couple of hundred yards north is another delightful scene, where on either side of the canal a few cottages, pretty gardens and moored boats face the waterway and the (very low) iron bridge. Then really tall, steep hills appear on the east bank, rising to over 250 feet. Three isolated cottages cling to the hill in a clearing among the trees. The canal leaves this damp, mossy area and bends round to Kinver lock. There is a pub here, and a road leading round to the village, which is behind the bold modern (1939) waterworks. This pumps from the vast underground lake that lies deep below the great sandstone ridge stretching from Kidderminster to Wombourn. Upstream of Kinver lock are the Kinver Cruiser Club moorings and the Dawncraft boatyard. At the next lock (Hyde lock) is another fine cottage: some canal wares may be bought here, as well as excellent home-made provisions—cakes, scones and bramble jelly. Lunch, tea and supper are served if prior notice is given (Kinver 3244). Beyond the lock, the canal wanders along the edge of woods on the east side of the valley, passing through the charmingly diminutive (25 yards long) Dunsley Tunnel, a rough-hewn bore with overhanging foliage at each end. The next lock is at Stewponey, where Stourton Castle stands just over the river. The Stourbridge Canal leaves at Stourton Junction, north of the wharf; the first of the many locks that carry the canal up towards Dudley and the Birmingham Canal Navigations is just a few yards away. (*See page 98*). Beyond Stourton Junction, a 90-degree bend to the left takes the canal to an aqueduct over the river Stour: this river now disappears to the north east and is not seen again. Its place near the canal is taken by the little Smestow Brook as far as Swindon.

At the far end of the aqueduct is a curious narrow boathouse (known as the Devil's Den) cut into the rock. Prestwood Park is concealed in the woods above the east bank. The hall is now a hospital: it used to be the home of the Foleys, a family of Black Country ironmasters. Meanwhile the canal continues through Gothersley and Rocky locks, reaching eventually a fork: the narrow entrance on the right leads into the very long Ashwood Basin, where there is a boatyard.

Ashwood Basin
This used to be a railway-connected basin owned by the National Coal Board. After the line was closed, the basin was disused for some years; but now it provides a pleasant mooring site for a large number of pleasure boats. There is a boatyard and a boatclub here. A road is carried over the basin by a small viaduct.

Stewponey Wharf
An interesting wharf at the head of Stewponey lock. The octagonal toll office has been recently restored by the boatyard that is based here. Near the wharf can be seen the long-abandoned track of the Kinver Valley Light Railway, which used to run from Stourbridge to Kinver. From Stewponey to Kinver it followed a route close to the canal. Just across the river from the wharf is the impressive bulk of Stourton Castle (see below), while in the opposite direction—but shielded by trees—is a fast main road, a built up area and a large roadhouse serving food. There is a

telephone box and a small grocery hard by. A petrol station is not far away.

Stourton Castle

Just a few yards west of Stewponey Wharf, this building is a curious mixture of building styles and materials. The castle is notable as the birthplace of Cardinal Pole in 1500. A friend of Mary Tudor, Pole became Archbishop of Canterbury in her reign after Cranmer had been burned at the stake. The Castle is private—it belongs to a former director of the Staffordshire & Worcestershire Canal Company—but visitors may be allowed to look round if they make an appointment (Kinver 2822).

Kinver

Staffs. PO, tel, stores, garage, bank.

Kinver clearly has a reputation as a Very Pretty Village. It is surrounded by tall hills and consists of a long main street of reasonably attractive houses, but its chief glory is its situation—it nestles among tall wooded hills, a position that must strike the visitor as remarkable for a village so close to the industries and unexciting geographical features of the West Midlands. Kinver Edge, (National Trust property), west of the village, is a tremendous ridge covered in gorse and heather, and for anyone prepared to toil up to the top from the valley it provides a splendid view of the Cotswold and Malvern hills. Kinver church is near the Edge, and is reached by a steep zig-zag road. The church overlooks the village and contains several things of interest, including plaques recording the Charter granted by Charles I in 1629 and the Charter granted by Ethelbad in 736, giving '10 cessapis of land to my general Cyniberte for a religious house'.

BOATYARDS & BWB

Ⓑ **Renown Marine Services** Ashwood Marina, Greensforge, Kingswinford, Staffs. (Kingswinford 79527). Canalside petrol and diesel. Chandlery. New boats sold. Boat repairs, inboard & outboard engine installation and repairs.

Ⓑ **Dawncraft** Stewponey lock, Stourton, near Stourbridge, Worcs. (Kinver 2481). Operational from March 1973. Sales of fibreglass ski and sports boats. Water, chandlery. Outboard engine servicing. *Open daily throughout the year.*

Ⓑ **Dawncraft** The Paddock, Kinver, Staffs. (2363). Water, canalside petrol, chandlery, dustbins, sewage disposal. Slipways, moorings, winter storage. Boat building, sales and repairs (all in fibre glass boats). Inboard & outboard engine repairs. *Open all year.*

BOAT TRIPS

Narrowboat 'Bellatrix', a converted trading boat built in 1935, runs trips along the canal from Kinver lock. Food and drink can be laid on; maximum number 45 passengers. Party bookings only. Enquiries to The Midland Navigation Packet Boat Service, 47 Geneva road, Shrubbery Estate, Tipton, Staffs. (021 557 7347).

PUBS

Stewponey & Foley Arms 50yds E of Stewponey lock. Steak house: meals served *daily except Sun, 12.30–14.15 and 19.00–22.15. On Suns, lunch only is served from 13.45–14.15,* and is usually a traditional roast joint. (Kinver 2835).

Vine by Kinver lock.

Ye Old White Hart Kinver.

Whittington 300yds E of bridge 28, along a footpath. Dating from 1300, this pub has an impressive history. It was the home of Dick Whittington's grandfather, and much later of Lady Jane Grey, whose ghost is sometimes encountered in the inn. There are also priest holes and a tunnel to the nearby Whittington Hall. Restaurant meals served here *daily except Sun, 12.00–14.00 and 19.00–22.00.* Reservations to Kinver 2110.

Anchor 200yds W of bridge 28. 15thC residential inn—lunches and dinners *daily throughout the year.* No public bar. (Kinver 2085 and 3291).

Bratch locks after a facelift.

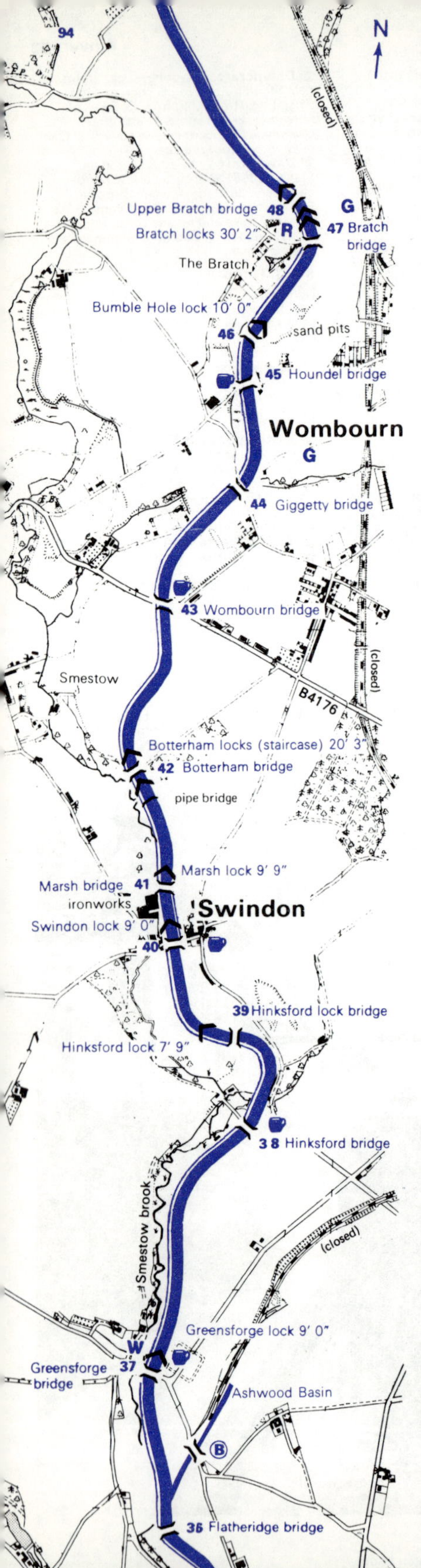

The Bratch

5 miles

North of Rocky lock, the outcrops of sandstone are seen less frequently, the countryside becoming flatter and more regular. However the locks do not disappear, for the canal continues the steady rise up the small valley of the Smestow Brook through southern Staffordshire towards Wolverhampton. At Greensforge lock there is an attractive pub, and another of the pretty, circular weirs that are found, often hidden behind a wall or a hedge, at many of the locks along this delightful canal. Another wooded, rocky section leads again to more open country. At Hinksford there is a residential caravan site down in the valley, and another pub near bridge 38, with a public telephone outside. On through the isolated Hinksford lock, the navigation bends round to Swindon, where a small ironworks adjoins the canal. Strengthening girders under bridge 40 have reduced its headroom to 6' 6". There are 4 locks hereabouts, Botterham being a 2-step staircase with a bridge crossing in the middle. North of here, the canal begins to lose its rural character as it encounters the modern outskirts of Wombourn, passing under a new bridge that, happily, retains the original cast-iron name plates. There is a pub by this bridge (43) and a few shops not far to the north east of the next one (44). Yet another pub is at the next bridge (45); beyond it is Bumble Hole lock, near the old gravel quarries. The 3 Bratch locks are next, raising the canal level by over 30 feet.

Boatmen may at first be completely foxed by these unique locks, but in fact they are much less complicated than they look (*see below*). From the top of the locks, one may enjoy a good view back down the valley, with the spire of Wombourn church backed by the great ridge of the Orton Hills to the east. At the foot of the locks is the great florid Bilston Waterworks, built in 1895. Leaving the Bratch, the canal wanders through open farmland parallel to a long-closed railway, arriving at the pleasantly situated Awbridge lock, with its circular weir and farm beside it.

Bratch locks
With their octagonal toll office, attractive situation and very unusual layout, these 3 locks are well known among students of canal engineering and architecture. At first sight they appear curiously illogical, with an impossibly short pound between the bottom of one lock and the top gate of the next; but the secret of their operation is the side ponds hidden behind the towpath hedge and the culverts that connect these to the intermediate pounds. In fact, to work through these locks, the boatman should simply treat each one as a separate lock like any other on the canal. *However, it is especially important when locking through to close the gates and paddles of each lock before operating the paddles of the next.* The big mistake is to consider the locks as a staircase—this will inevitably lead to things going wrong.

Wombourn
Staffs. PO, tel, stores. This town used to be an attractive village, before an expansion scheme flooded the town with acres of council houses.

Swindon
Staffs. PO, tel, store, fish & chips. A small village, a mixture of farming and industry. The farming was obviously here first, the old buildings forming the core of the village, but the 19thC ironworks determine the nature of the canal's surroundings.

PUBS

- **Round Oak** Canalside, at bridge 45.
- **Waggon & Horses** Near the canal at bridge 43, on B4176.
- **Green Man** Swindon Ironworks, 100yds W of bridge 40.
- **Old Bush Inn** Swindon village, 150yds E of bridge 40.
- **Old Bush** Hinksford, 100yds NE of bridge 38. Garden.
- **Navigation** Greensforge. Canalside, at the lock. Snacks and local draught beer.

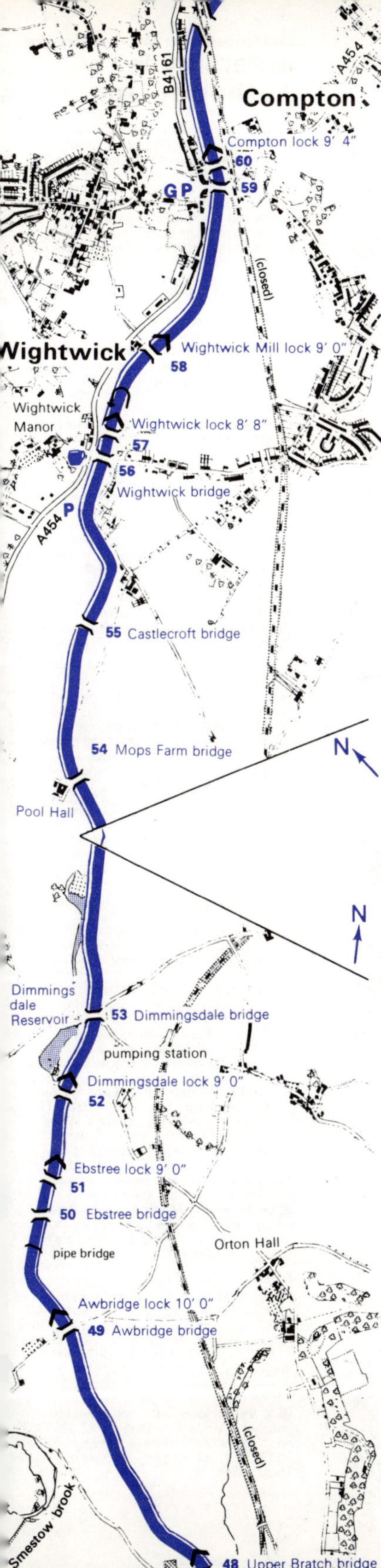

Dimmingsdale

4¾ miles

North of Wombourn, the countryside becomes less interesting. Although it is still quiet, and remote from roads and railways, the canal is pestered by overhead power lines for a mile as it rises through Awbridge, Ebstree and Dimmingsdale locks. The modest, pretty lakes alongside here are in fact canal-feeding reservoirs, of considerable interest to fishermen. Ahead, the hills of Wightwick overlook the navigation as it approaches a shallow valley and a busy main road. This valley–in places an artificial cutting–contains the canal right through to the flatter land at Autherley. Houses, mostly modern, are never far away, although the canal manages to preserve intact its rural character all the way through what are in effect the western outskirts of Wolverhampton. Compton lock marks the end of the 31-lock climb from the river Severn at Stourport, a rise of 294 feet. From here northwards, a 10-mile level pound takes the Staffs & Worcs on to Gailey, where the first of 12 locks begins the fall towards the Trent & Mersey Canal.

Compton
Staffs. PO, tel, stores, launderette. A busy but uninteresting village with a modern shopping centre and pubs that are less than picturesque. The canal lock was the first that James Brindley built on the Staffs & Worcs in the late 1760's, but unfortunately the cottage here was recently demolished. However the nearby coal wharf still handles coal, although of course the waterway is no longer used for this trade–nor for any other.

Wightwick
Staffs. (pronounced 'Wittick'). Once a village, now a suburb. But there is a hill and plenty of trees, and in spite of the busy A454 it is a pleasant scene around the canal bridge and the old country pub.

Wightwick Manor 300yds NW of bridge 56, across the A454 and up the hill. Built between 1887 and 1893, the Manor has an exterior that embodies many of the idiosyncrasies of the time. Inside, it is furnished with original wallpapers and fabrics by Morris and various contributions by the Pre-Raphaelites. This certainly makes a change from the usual venerable stone or timbered buildings that are open to the public. National Trust property. *Open all year 14.30–16.30 on Thur, Sat and B. Hol weekends, also Wed from May to Sep.* (Wolverhampton 761025).

BOATYARDS & BWB

Ⓑ **Mermaid Hire Cruisers** Wightwick Wharf, Castlecroft, Wolverhampton (763818). By bridge 56. Hire cruisers, engine and boat repairs. Groceries and fuels nearby.

PUBS

Oddfellows Hall Compton, 50yds W of bridge 59.

Swan Compton, near the Oddfellows.

Mermaid Wightwick, 50yds NW of bridge 56. *Lunches Mon-Sat, 12.00–14.00.* Sandwiches usually obtainable. (Wolverhampton 761395).

FISHING THE STAFFORDSHIRE & WORCESTERSHIRE CANAL

There is good coarse fishing throughout the length of this canal. There is a wide variety of species well distributed, notably bream, roach, perch and pike. Among the most popular reaches are from Cookley to Kinver, including Kidderminster and Stourport–held by the Birmingham AA, whose membership includes all this fishing. Local anglers fish the canal regularly and catch roach up to just over a pound in weight. The canal has been restocked in several places, and roach fishing is now improving following a shortage of these fish. There are also carp in places.

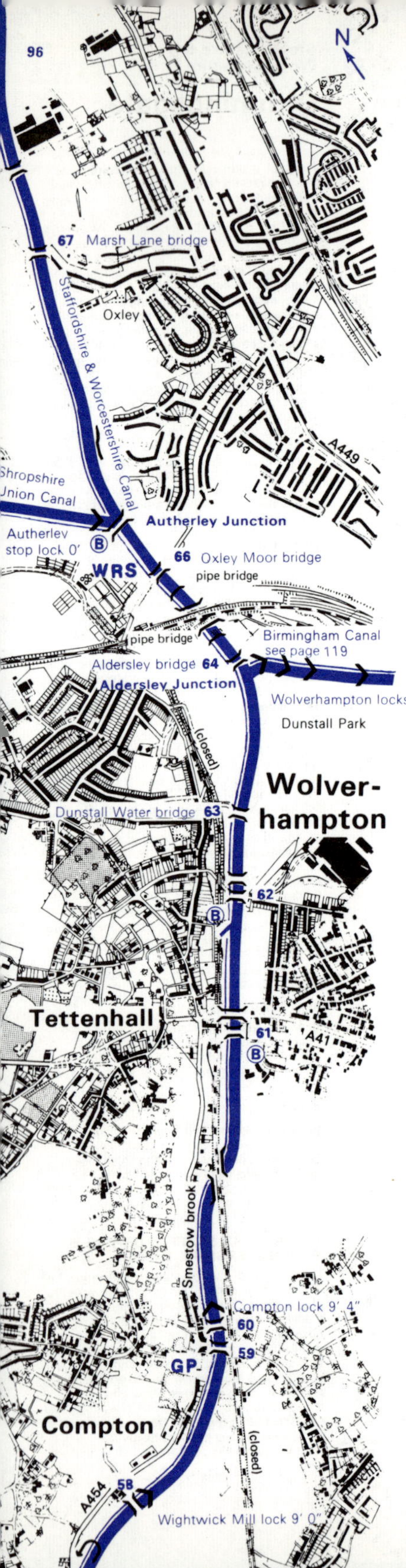

Staffordshire & Worcestershire

Aldersley & Autherley Junctions

The canal continues along the pleasant cutting through the Wolverhampton outskirts, passing under the big iron bridge that carries a closed railway. Trees on either side shield the navigation from the houses of Tettenhall. A large concrete road bridge that dwarfs the nearby original brick-arched bridge reminds the traveller of the hurly-burly world on either side of this quiet corridor of green. Several boating centres are passed, then the trees of Dunstall Park appear on the right, followed by Aldersley Junction. At this secluded place the bottom of the 21 Wolverhampton locks brings the main line of the Birmingham Canal into the Staffordshire & Worcestershire Canal. There is a towpath 'roving' bridge at the junction; past it is a criss-cross of railway and pipe bridges, then the cutting eases away and the canal goes under a road bridge, emerging at last in a rather flat, open landscape, with council housing estates to the east. The big white bridge on the towpath side marks Autherley Junction: the stop lock here is at the entrance to the former Birmingham & Liverpool Junction Canal, now part of the Shropshire Union main line. There is a large boatyard north of the stop lock, and extensive BWB moorings. Straight ahead at Autherley is the continuation of the Staffs & Worcs to the Trent & Mersey Canal at Great Haywood Junction. *All these canals are covered in volume 2 of this series, the Guide to the Waterways-North West. For coverage of the Birmingham Canal main line, see page 116 below.*

Tettenhall
Staffs. A comfortable residential suburb of Wolverhampton (the city centre is a mere 2 miles to the south east). There is little to see, but the battlemented tower of the Norman church is worth a look–it is 300yds N of bridge 61). The church was burnt down in 1950, only the tower surviving the conflagration. The new church, built in 1955, contains some interesting stained glass. PO, tel, stores, garage, pub and other facilities are conveniently placed by bridge 61.

BOATYARDS & BWB

Ⓑ **Water Travel** Autherley Junction, Tettenhall, Wolverhampton, Staffs. (Fordhouses 2371). On the Shropshire Union canal: used to be Melody Line Cruisers. Water, diesel, petrol, gas, chandlery. Groceries and canal ware. Slipway, moorings. Boat building, sales and repairs, engine sales and repairs. The adjoining Autherley Boat Club welcomes visitors from the canal to its licensed clubhouse. Full length converted narrowboat available for charter by parties for short trips–bar on board.

Ⓑ **Double Pennant Boatyard** Hordern Road, Wolverhampton, Staffs. (752771). Just S of bridge 62. Hire cruisers. Water, diesel, petrol, gas, chandlery, clothing. Dustbins and sewage disposal. Slipway. Boat repairs and fitting out. Inboard & outboard engine sales and repairs: agents for Chrysler, Crescent and Penta.

Ⓑ **Leisurecraft Marine** Newbridge Wharf, Tettenhall Road, Wolverhampton, Staffs. (752368). Fibreglass boat building & sales. By bridge 61.

BOAT TRIPS

Narrowboat 'Compton Queen' runs trips from Autherley carrying parties of up to 40. Also hire cruisers and moorings at Autherley (on the Staffs & Worcs canal). All enquiries to Mr Gregory, 83 Aldersley rd, Tettenhall, Wolverhampton, Staffs. (753851).

PUBS

Newbridge Tettenhall, 50yds E of bridge 61. Meals here *Mon-Fri 12.00–14.00 and 19.00–21.00.* (Wolverhampton 754911). The inn sign is a picture of Thomas Telford.

Stourbridge

Maximum dimensions

Length: 70′
Beam: 7′
Headroom: 6′

Mileage

STOURTON JUNCTION to
Wordsley Junction: 2
Stourbridge: $3\frac{1}{4}$
BLACK DELPH bottom lock: $5\frac{1}{4}$

Total 16 locks

The Stourbridge and Dudley Canals are to some extent inseparable, being part of the same grand scheme to link the Dudley coal mines with the Stourbridge glass works, and with the Severn navigation by means of the Staffs & Worcs Canal. The Acts for the two canals were passed on the same day in 1776. Well supported by local glass masters, the Stourbridge Canal was soon under way, with Thomas Dadford as engineer. From a junction with the Staffs & Worcs at Stourton, the canal ran to Stourbridge. There was a 2 mile branch to the feeder reservoirs on Pensnett Chase, with 16 rising locks, and another level branch ran to Black Delph, where it met the Dudley Canal.

Combined with the Dudley Canal, the Stourbridge was soon profitable, and it was not long before the two companies were seeking to increase their revenue. They decided to try to capture some of the rich traffic on the Birmingham Canal. A joint proposal for a junction with the Birmingham Canal line via the Dudley tunnel was authorised in 1785, and the through route was opened in 1792. Although the two companies worked so closely together, they resisted the temptation to amalgamate, relying on their mutual dependence to sort out any problems. Although the Dudley Canal became part of the BCN system the central position of the Stourbridge made it a very profitable undertaking, and throughout the early years of the 19thC the trade steadily increased. Even the opening of the rival Worcester & Birmingham Canal in 1815 did not affect the profits. Revenues were further increased in 1840 when the Stourbridge Extension Canal, later GWR owned, was opened to capture the coal trade from the Shut End collieries.

In the middle of the 19thC railway competition began to affect the canal, first from the Oxford Worcester & Wolverhampton Railway, and later from the Great Western Railway. Revenues began their inevitable decline, but the Stourbridge was able to maintain its profits, and thus its independence. It never merged with any other company, nor was it bought by any railway; it was able to retain its independence until nationalisation in 1948, although by then it was in the same run down state as most canals in Britain. Commercial traffic died away, and by the 1950s the canal was no longer usable.

In 1964 the Staffs & Worcs Canal Society and the BWB decided to restore the 16 locks, and re-open the line between Birmingham and the Severn. Using volunteer labour provided by the Society and money and know-how provided by BWB, work slowly progressed, and in 1967 the Stourbridge Canal was re-opened to traffic, with pleasure boats taking the place of the trading boats that had made the canal so profitable in earlier days.

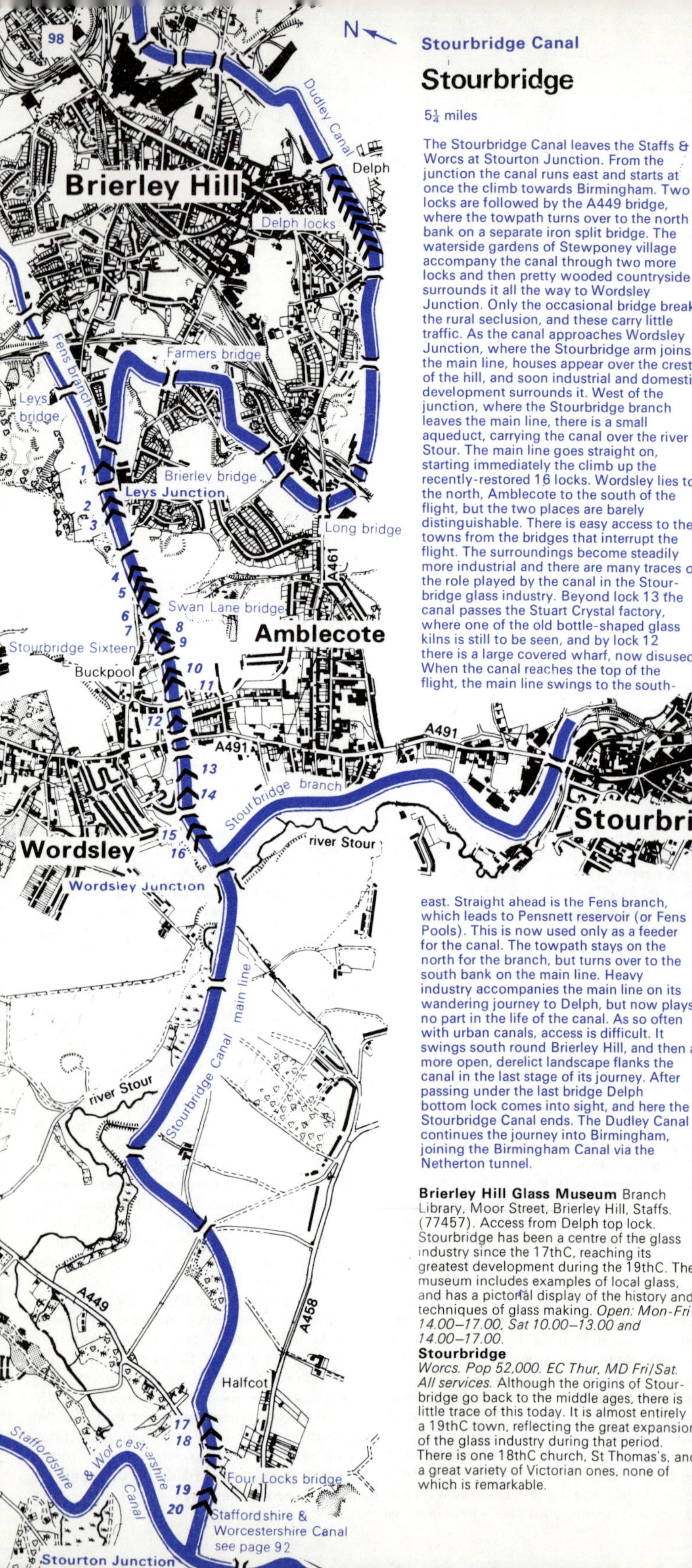

Stourbridge Canal

Stourbridge

$5\frac{1}{4}$ miles

The Stourbridge Canal leaves the Staffs & Worcs at Stourton Junction. From the junction the canal runs east and starts at once the climb towards Birmingham. Two locks are followed by the A449 bridge, where the towpath turns over to the north bank on a separate iron split bridge. The waterside gardens of Stewponey village accompany the canal through two more locks and then pretty wooded countryside surrounds it all the way to Wordsley Junction. Only the occasional bridge breaks the rural seclusion, and these carry little traffic. As the canal approaches Wordsley Junction, where the Stourbridge arm joins the main line, houses appear over the crest of the hill, and soon industrial and domestic development surrounds it. West of the junction, where the Stourbridge branch leaves the main line, there is a small aqueduct, carrying the canal over the river Stour. The main line goes straight on, starting immediately the climb up the recently-restored 16 locks. Wordsley lies to the north, Amblecote to the south of the flight, but the two places are barely distinguishable. There is easy access to the towns from the bridges that interrupt the flight. The surroundings become steadily more industrial and there are many traces of the role played by the canal in the Stourbridge glass industry. Beyond lock 13 the canal passes the Stuart Crystal factory, where one of the old bottle-shaped glass kilns is still to be seen, and by lock 12 there is a large covered wharf, now disused. When the canal reaches the top of the flight, the main line swings to the south-east. Straight ahead is the Fens branch, which leads to Pensnett reservoir (or Fens Pools). This is now used only as a feeder for the canal. The towpath stays on the north for the branch, but turns over to the south bank on the main line. Heavy industry accompanies the main line on its wandering journey to Delph, but now plays no part in the life of the canal. As so often with urban canals, access is difficult. It swings south round Brierley Hill, and then a more open, derelict landscape flanks the canal in the last stage of its journey. After passing under the last bridge Delph bottom lock comes into sight, and here the Stourbridge Canal ends. The Dudley Canal continues the journey into Birmingham, joining the Birmingham Canal via the Netherton tunnel.

Brierley Hill Glass Museum Branch Library, Moor Street, Brierley Hill, Staffs. (77457). Access from Delph top lock. Stourbridge has been a centre of the glass industry since the 17thC, reaching its greatest development during the 19thC. The museum includes examples of local glass, and has a pictorial display of the history and techniques of glass making. *Open: Mon-Fri 14.00–17.00, Sat 10.00–13.00 and 14.00–17.00.*

Stourbridge
Worcs. Pop 52,000. EC Thur, MD Fri/Sat. All services. Although the origins of Stourbridge go back to the middle ages, there is little trace of this today. It is almost entirely a 19thC town, reflecting the great expansion of the glass industry during that period. There is one 18thC church, St Thomas's, and a great variety of Victorian ones, none of which is remarkable.

Worcester & Birmingham

Maximum dimensions

Length: 71′6″
Beam: 7′
Headroom: 6′

Mileage

WORCESTER, Diglis Basin to
Tibberton: 5¾
Dunhampstead: 7½
Hanbury Wharf: 9¼
Stoke Wharf: 12¾
Tardebigge top lock: 15½
Bittell Reservoirs: 20½
KING'S NORTON JUNCTION: 24½
BIRMINGHAM Worcester Bar Basin: 30

Total 58 locks

The Bill for the Worcester & Birmingham Canal was passed in 1791 in spite of fierce opposition from the Staffordshire & Worcestershire Canal proprietors, who saw trade on their route to the Severn threatened. The supporters of the Bill claimed that the route from Birmingham and the Black Country towns would be much shorter, enabling traffic to avoid the then notorious shallows in the Severn below Stourport. The Birmingham Canal Company also opposed the Bill and succeeded in obtaining a clause preventing the new navigation from approaching within seven feet of their water. This resulted in the famous Worcester Bar separating the two canals in the centre of Birmingham.

Construction of the canal began at the Birmingham end following the line originally surveyed by John Snape and Josiah Clowes. Even at this early stage difficulties with water supply were encountered. The Company were obliged by the Act authorising the canal to safeguard water supplies to the mills on the streams south of Birmingham. To do this, and to supply water for the summit level, ten reservoirs were planned or constructed. The high cost of these engineering works led to a change of policy: instead of building a broad canal, the Company decided to build it with narrow locks, in order to save money in construction and water in operation.

Work on King's Norton Tunnel, described at the time as 'a stupendous undertaking', began in 1794; by 1807 boats could get from Birmingham to Tardebigge Wharf. Here work came to a standstill for several years while the Company considered alternative cheaper ways of completing the line down to the Severn. Work eventually started again under a new engineer, John Woodhouse, a great exponent of boat lifts. He proposed reducing the number of locks down to Worcester from 76 to 12, using lifts to descend most of the fall. The Company were less enthusiastic and limited his enterprises to one experimental lift at Tardebigge. This seems to have worked reasonably well but the Company were still sceptical. They called in the famous canal engineer John Rennie, who decided that the mechanism would not withstand the rough treatment that it would doubtless receive from the boatmen. Consequently locks were built but reduced in number to 58. The site of the lift became the top lock of the Tardebigge flight—which accounts for its unusual depth.

After this, work progressed steadily and the canal was completed in 1815. In the same year an agreement with the Birmingham Canal proprietors permitted the cutting of a stop-lock through the Worcester Bar. The canal had cost £610,000, thus exceeding its original estimate by many thousands of pounds. Industrial goods and coal were carried down to Worcester, often for onward shipping to Bristol, while grain, timber and agricultural produce were returned to the growing towns of the Midlands. The canal basins in Worcester became important warehousing and transhipment points: Diglis Basin had warehousing for general merchandise, grain and wine, and Lowesmoor Basin specialised in coal and timber. Prosperous businesses were conducted from these wharves and they were an important port of call for the main canal carriers. However the opening of railways in the area in the 1840s and 1850s reduced this traffic considerably and had a profound effect on the fortunes of the canal.

In an attempt to win back salt carrying

the Canal Company cut the Droitwich Junction Canal in 1852 to connect the Droitwich Barge Canal and the town of Droitwich with their main line at Hanbury Wharf. Toll income and profits continued their relentless decline, however, and after 1864 the Company was unable to pay a dividend. In 1874 the canal was bought by the Sharpness New Docks Company. (The words 'Sharpness New Docks and Gloucester & Birmingham Navigation Company' can still be seen on old notices on some of the bridges.) The new management commenced a programme of works to improve the canal in the hope of attracting trade but, in effect, the canal was subsidised by the Gloucester & Berkeley Ship Canal for the rest of its working life.

The animals that used to draw the boats along the Worcester & Birmingham Canal were mainly donkeys worked in pairs instead of the more usual horse. Why this should have been so is not recorded but both horse and donkey were unsatisfactory on the summit level with its four tunnels, for only the short Edgbaston Tunnel has a towpath through it. To overcome the delays caused by the need to 'leg' boats through the other tunnels, steam tugs were introduced in the 1870s and successfully hauled trains of boats through the tunnels for many years.

By the early 1900s the commercial future of the canal was uncertain, although the works were in much better condition than on many other canals. Schemes to enlarge the navigation as part of a Bristol-Birmingham route came to nothing. Commercial carrying continued until about 1964, the traffic being mostly between the two Cadbury factories at Bournville and Blackpole, and to Frampton on the Gloucester & Sharpness Canal. Now occasional coal boats are the only ones carrying goods to Worcester. After nationalisation several proposals were made to abandon the canal but the 1960s brought a dramatic increase in the number of pleasure boats using the waterway, and its future is now secure. Now it is part of the popular cruising circuit comprising the river Severn, the Staffordshire & Worcestershire Canal and part of the Birmingham Canal Navigations. Soon it will be part of another circuit when the Upper Avon Navigation is re-opened in 1973.

Natural history

The Worcester & Birmingham Canal provides a rich variety of plant, bird and insect life throughout its length. Apart from its general interest it provides material for more serious study since it flows past extensive salt deposits left by an ancient sea of the Triassic period, consequently the water is quite brackish in some localities. After emerging from the gloom of the West Hill Tunnel it is worth pausing at the exit not only to admire the tree-covered avenue but to examine the variety of ferns and liverworts which grow on the dripping clay of the high banks. A mile or so downstream, a mooring at Bittell is a must, for the reservoirs are among the best-known bird haunts in this part of the country, particularly for wild duck. One large reservoir (Bittell Lower) lies immediately against the west bank of the canal. There are always great crested grebe and mallard here, but the commonest water bird is the coot.

The Tardebigge flight of locks gives splendid views over open meadow land where the song of the skylark is usually heard in summer. Here the clumps of great willow herb and orange balsam line the banks. The orange balsam was introduced from America, being first recorded in Surrey in 1822; it has now spread along the river and canal system to most of England. When touched, the pods shoot out their seeds to a distance of several feet.

Where the vegetation is dense the harsh scolding of the sedge warbler can be heard; it is particularly frequent in this reach of the canal where caddis flies and other insects are abundant. Two common dragonflies are the large yellow aeshna and the delicate damselfly ischura elegans, which has a vivid blue spot on the abdomen. When the locks are emptied the walls can be seen to be covered with thousands of aquatic snails (limnaea). With such an abundance of aquatic organisms it is not surprising that the Worcester & Birmingham Canal is a favourite among anglers. There are roach and plenty of 3lb bream; and a 6lb carp has been recorded.

The calls of redshank and common sandpiper are not unfamiliar at Tardebigge for these waders often fly over on their way to Tardebigge reservoir which is beside the locks. In addition to waders and waterbirds herons are usually to be seen feeding in the shallows. In late summer varieties such as the wood sandpiper and the black tern may be occasionally seen at both the Bittell and Tardebigge reservoirs.

At Stoke Prior a subtle change comes over the canal, for the water becomes rather brackish. One interesting waterweed is the enteromorpha—a kind of alga which looks like floating transparent tubes about half an inch across. Although the canal is enclosed by brick walls for a short distance, a mass of rosebay willow herb clothes the industrial scars in summer and parts of the waterway are covered with yellow water lilies.

From the Astwood flight to past Hanbury there are many beds of phragmites (reeds) on the west bank. Reed warblers are common in this area as they can only weave their suspended nests in the stems of these plants. Moorhens are common throughout the canal and, in places where bushes and trees hang into the water, families of mallard can usually be seen. Where the bank is low, a grass snake may sometimes be observed swimming in the water hunting for frogs or small fish. The water vole is common everywhere. Being rather short-sighted and slightly deaf it can sometimes be watched from a distance of three feet for, like many creatures of the canal, it expects intruders only from the towpath.

Enquiries concerning natural history may be addressed to the Department of Natural History, the City Museum and Art Gallery, Birmingham, B33DH.

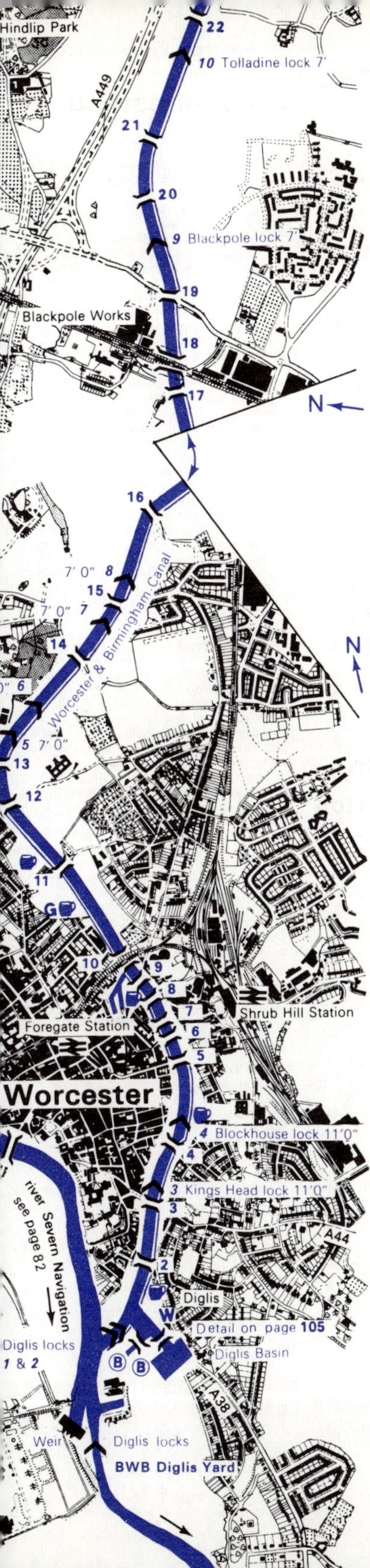

Worcester

4 miles

The Worcester & Birmingham Canal begins at Diglis, on the south side of Worcester. It leaves the river Severn a few hundred yards north of the Diglis locks, climbs 2 wide (18ft) locks and opens out into one of the two Diglis basins, where a large number of pleasure boats are moored, several of them sea-going: a couple of boatyards and other boating facilities are available here. Past the basin, the canal becomes hemmed in by the town as it enters the first of the many deep, narrow locks up to Birmingham. The canal curves round the east side of Worcester, between the town and Shrub Hill station. Access to the town centre is possible at most of the bridges, although the canal is virtually ignored by the town—it is lined by old buildings and intermittent waste ground, and hardly anywhere is it consciously used as a positive part of the urban scene. The only centre of canal activity is at Diglis. At one point, near the railway viaduct that leads to Foregate station and the west, the towpath rises over the entrance to the old Lowesmoor basin and wharves. These are unused and inaccessible, but may soon be brought back into use as canal boat moorings. A series of four locks lifts the canal up and away from the outskirts, then it is crossed by a railway line, with an isolated industrial works beside it. The canal turns east for several miles, negotiating Blackpole lock and then Tolladine lock, with its pretty cottage and garden. Worcester is well behind, and the canal is in open country, but the A449 runs parallel. Hindlip Park is to the north.

Worcester
For notes on Worcester, *see page 82*.

Diglis Basin
This is a fascinating terminus at the junction of the river Severn and the Worcester & Birmingham Canal. It consists of basins, boatyards, old warehouses and a drydock. Commercial craft have been entirely replaced by a mixture of pleasure boats designed for narrow canals, rivers and the sea. (The latter predominate, appreciating the sheltered moorings in deep water off the river Severn.) There are plenty of facilities, mostly catering for the sea-going boats: 2 boatyards, a chandlery, and the usual BWB facilities. Permission to use the drydock should be sought from the BWB basin attendant, whose house is at the top of the 2 locks down into the river: *telephone Worcester* 26235. The locks will take boats up to 72ft by 18ft 6in, although obviously only narrow boats can proceed along the canal beyond the first lock. The locks (which can be operated only by the basin attendant, between 8.00 and 19.00) incorporate a side pond to save water; and near the second lock is a small pump house that raises water from the river to maintain the level in the basin.

BOATYARDS & BWB

Ⓑ **J. H. Everton Ltd** Diglis Basin, Worcester (20973). Boat builders & repairers, marine engineers. 10-ton gantry crane. Moorings (some covered).

Ⓑ **Cathedral Cruisers** Basin road, Diglis, Worcester (22021 or 820691). Boat building and repairing. Canal and river hire cruisers.

BOAT TRIPS

Worcester Steamer Company Diglis, Worcester (27543). Large motor trip boats operating along river Severn, for party bookings (up to 200 passengers carried). Also public service in summer. Buffet and music can be arranged.

PUBS

Cavalier Tavern Worcester. Canalside, at bridge 11.

Bridge Worcester. Canalside, at bridge 9.

Bricklayers Arms Worcester. 10yds from Blockhouse lock.

Anchor Diglis, on the road just outside the basin grounds.

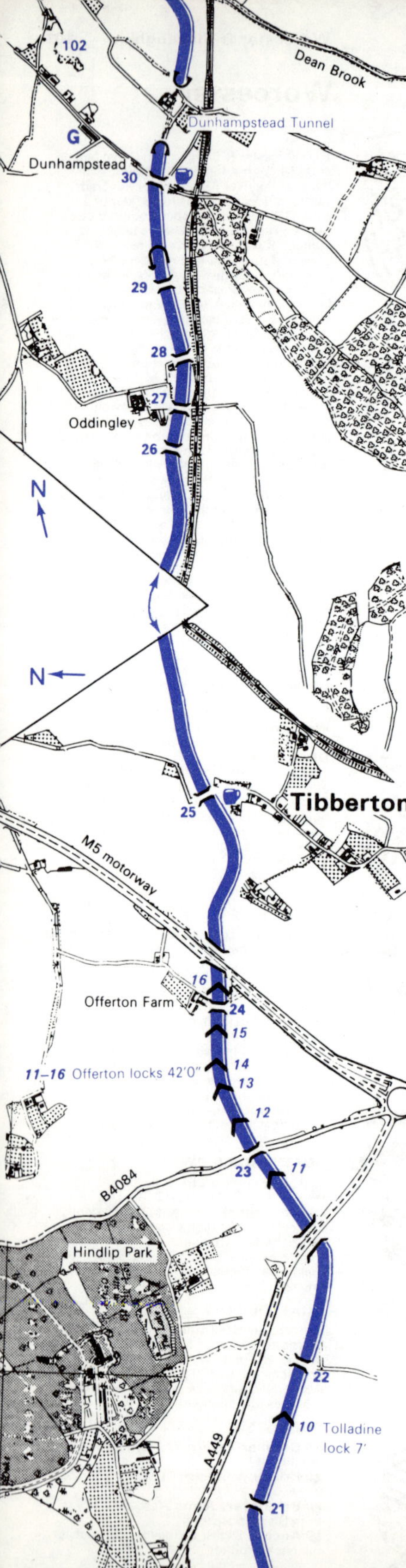

Dunhampstead

4½ miles

This is a very pleasant stretch of rural canal, entirely typical of the Worcester & Birmingham. Having left the outskirts of Worcester, and accompanied by a minor road, the canal now goes under the A449 and ascends the six Offerton locks, which are set in pleasant pastureland with a pretty cottage near the top. The M5 motorway crosses on its skewed steel bridge; it vanishes immediately as the canal enters a short, curving cutting that brings one to the village of Tibberton. There are 2 pubs nearby, and fruit trees remind one of Worcestershire's orchards. The canal moves towards a ridge of hills to the east, but a railway line intercedes to prevent the canal reaching the side of the valley. This is the main Bristol-Birmingham line, which carries many fast passenger trains, but the only station near the canal is at Bromsgrove. At Oddingley there is a little church and a timbered farm that look out together over the canal; further on are the crowded moorings at Dunhampstead Wharf (popular pub nearby). A wooded cutting leads to Dunhampstead Tunnel, the first of five between here and Birmingham. There is no towpath in the tunnel; horses used to walk over the hill while boatmen pulled the boats through by the handrail (still in place) along each side of the tunnel. It is 236yds long.

Dunhampstead
Worcs. PO, tel, stores (the shop is a few hundred yards NW of bridge 30). A hamlet consisting of no more than five buildings, including the railway signal box. The only life in the area, apart from the trains that roar past the woods, is provided by the canal and the nearby pub. There are moorings here, a narrowboat that runs trips, and a shop (previously a forge) selling canal wares and antiques. To the north is the tunnel.

Tibberton
Worcs. PO, tel, stores (all to the south of the pubs). A small but expanding canalside village, of little interest. There is a fine old rectory by the Victorian church.

BOAT TRIPS

The narrowboat 'Cedar' operates trips from Dunhampstead Wharf (bridge 30). Private party hire – parties up to 40 strong catered for, and bar & food available. Also regular summer Sunday service at 16.00 from Dunhampstead to Hanbury Wharf. Moorings available at Dunhampstead. All enquiries to Droitwich 3889.

PUBS

Fir Tree Dunhampstead, near bridge 30. Popular pub in pleasant surroundings. Garden.

Bridge Tibberton, near bridge 25. Garden.

God Speed the Plough Tibberton, near the bridge.

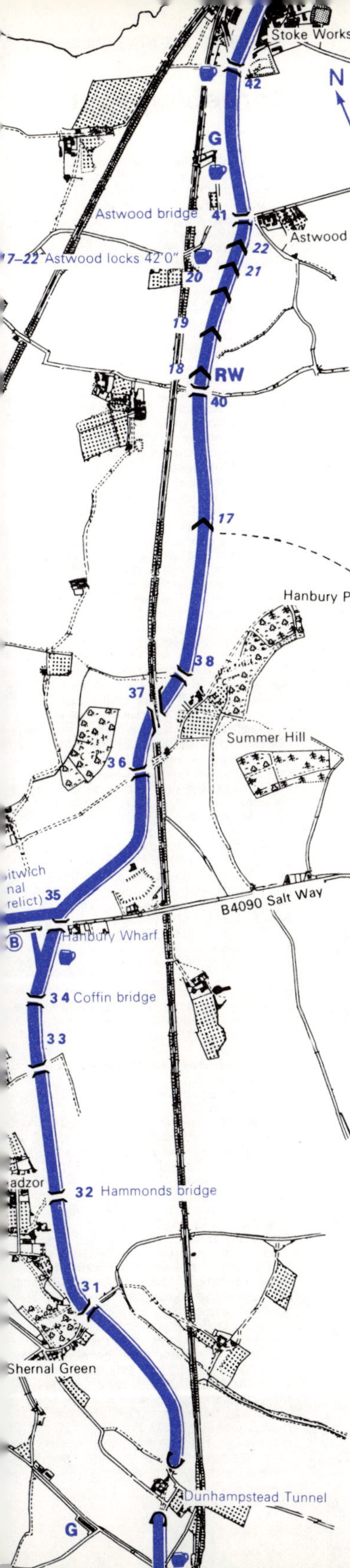

Hanbury Wharf

4⅝ miles

Leaving Dunhampstead Tunnel, the canal enters flatter countryside as the hills recede to the east. The pretty, residential settlement of Shernal Green flanks the canal, while Hadzor House (late 18thC in the classic tradition) is visible in the trees on the west side of the canal. This straight stretch is terminated by the very busy area of Hanbury Wharf, where an old arm and a new building comprise a boatyard for small pleasure boats. There is a pub by the main road bridge—the Droitwich Junction canal (now unnavigable) joins here. North of here the ridge of hills approaches again from the east. The boatman should relish this 5½ mile level—it is easily the longest pound between Worcester and Tardebigge top lock. But as the canal passes under the railway to take up an uninterrupted position on its east side, the ridge of hills nears again, accompanied by attractive parkland. Ahead are the 6 locks in the Astwood flight, set in pleasant open pastureland. Near the top is a pub, beside the railway line; and beyond the cottages at Astwood bridge is a semi built-up area—a minor road joins the canal, lined by workmen's terraced cottages. A useful grocery store and post office is here, as well as 2 more pubs, one of these having a verandah fronting the canal. To the north of this settlement is the reason for its existence—a huge industrial chemical works, owned of course by ICI. The canal goes through the middle of this works, much of which is derelict now.

Stoke Works
Now closed, this establishment was built in 1828 to pump brine (salt) from underground sources for industrial uses, and provided much of the canal's trade (later gained by the railways). The blackened works seems very out of place in this rural landscape and is, happily, the only outpost of industry on the canal between Worcester and King's Norton, near Birmingham.

Hanbury Hall
NT property. (Access via the public footpath leading SE from lock 17). Set in a spacious and well-wooded park, this is a Wren-style red brick house built in 1701 and little altered since then. On show are the long room and main staircase with painted ceilings by Thornhill. *Open Apr-Sep: Wed and Sat 14.00-18.00.*

Hanbury Wharf
An interesting canal settlement at the junction of the Droitwich Junction and Worcester & Birmingham canals. A short arm leads to the original wharf, but the old canal cottages here are now overshadowed by the big new shed that houses a busy modern boatyard. Construction of the Droitwich Junction canal from here down to Droitwich (2 miles to the west), and of the Droitwich canal on to the river Severn, presumably lessened the usefulness of Hanbury Wharf. The Droitwich Junction canal is disused and in places totally overgrown now—one can hardly recognise the junction just north of bridge 35—but there are plans afoot to reopen the length from the junction to the first lock as moorings.

BOATYARDS & BWB

Ⓑ **Ladyline (Hanbury Marina)** Hanbury rd, Droitwich, Worcs. (3002). Boat sales & repairs, inboard & outboard engine sales & repairs, fitting out. Slipway, permanent moorings. Water, gas, chandlery, rubbish & sewage disposal.

PUBS

Boat and Railway Canalside, just south of bridge 42. Terrace on to canal; food shops nearby.

Butcher's Arms near canal north of bridge 41; but access from the canal is difficult.

Bowling Green beside railway bridge, 200yds SW of bridge 41.

Eagle & Sun Hanbury Wharf. Canalside, at bridge 34.

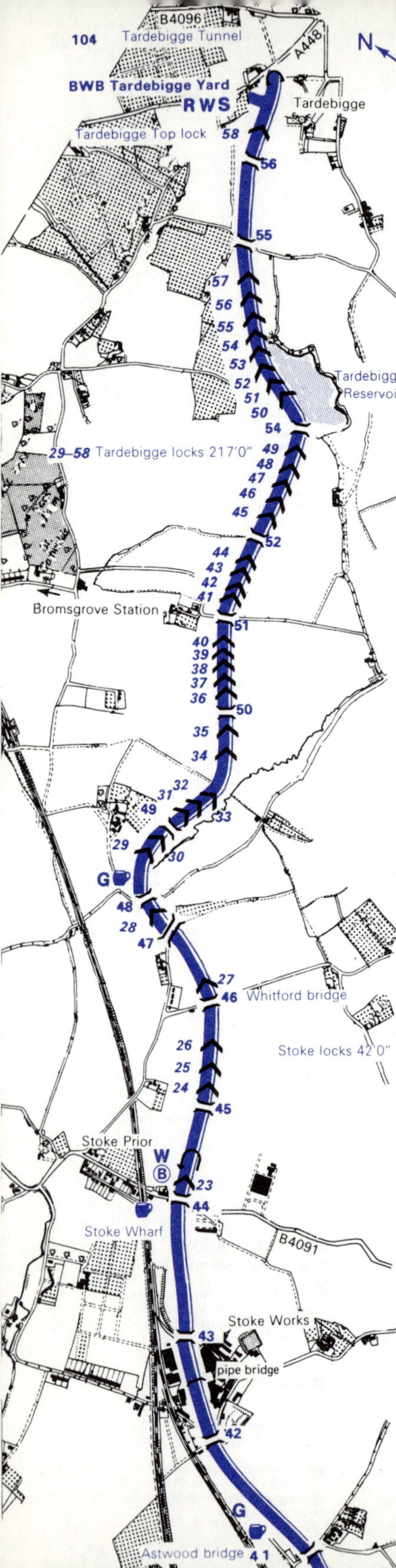

Tardebigge Locks

4¾ miles

Approaching Stoke Wharf from the old works, boatmen will note the hills to the north-east with some misgiving. At Stoke Wharf is the first lock for over a mile; beyond is a crowded mooring site, and then more locks, flanked by trees and pasture land. These locks (numbers 23–28) form the Stoke flight, but in fact there is only a short breathing space of a few hundred yards, with a well-placed pub, before the first of the 30 Tardebigge locks is reached. Forward progress becomes a crawl as this great flight is climbed, but the pleasures of the surroundings make the effort worthwhile. The locks wind up through pretty, folding countryside, leaving the busy railway behind in the west. There are attractive, well cared-for cottages scattered along the flight, generally near the bridges; their gardens overlook the canal. The locks themselves have great charm, being equipped throughout with traditional wooden gates and balance beams. Large paddles speed up locking, and so a reasonably well co-ordinated crew of two can work through a lock every 5 minutes. The remote rural course of the canal takes it well wide of Bromsgrove, but Bromsgrove station is only a mile north-west of bridge 51. Between locks 50 and 54 Tardebigge Reservoir can be seen behind an embankment on the east bank. This feeder reservoir is particularly popular with fishermen; the angling rights are let to the private Bournville Athletic Club. As the reservoir is about 50ft. below the summit level, a steam engine was installed to pump water up the hill. The engine-house still stands near the canal, but the engine was removed some time ago. Tardebigge top lock has a fall of 14ft., one of the deepest narrow locks in the country. When the canal was built, there was a vertical boat lift here. The usual technical problems caused the lift to be replaced by the deep lock, and there is little trace of it now. Above the lock is Tardebigge wharf, overlooked by the elegant spire of Tardebigge church, up on the hill to the east. At the wharf is a BWB maintenance yard, and a large mooring site. Leaving the wharf and its cottages behind, the canal vanishes into a tunnel, passing under a main road (A448) at the tunnel mouth.

Tardebigge
Worcs. Stores (on A448 10 mins walk from south end of Tardebigge tunnel). A small farming village flanking the main road. Apart from the settlement near the canal, the best part of the village is up on the hill, around the fine 18thC church with its delicate spire.

Avoncroft Museum of Buildings, Stoke Prior
1 mile N of Stoke Wharf, off B4091. (Bromsgrove 31363). Old buildings rescued from demolition are re-erected and displayed here. Exhibits include an 18thC post mill, a local nail and chain works, a 15thC timber-framed house from Bromsgrove, and the 14thC roof of Guesten Hall, Worcester. There is also a reconstruction of an Iron Age hut. *Open mid Mar-mid Oct, Tue-Sun 11.00-18.00.*

Stoke Wharf
A pretty canal settlement in the best tradition—a lock, a wharf and warehouse, a crane overhanging the water, a pleasant line of creeper-clad houses facing the canal. The boats moored here are mostly old narrowboats. Stoke Wharf is the only compact element of Stoke Prior—perhaps the heart of the village was drawn to the canal when the latter was built, and has remained there ever since. Stoke Prior church, which is mainly of the 12thC, stands by itself half a mile north of the wharf, the other side of the busy railway junction. Stoke Prior is not a good place for shopping: it is better to victual up at the settlement near bridge 42. However there is a pub near the wharf.

BOATYARDS & BWB

BWB Tardebigge Yard Tardebigge top lock. (Bromsgrove 72572). Water, dustbins and sewage disposal. Drydock available, also permanent moorings.

Ⓑ **Stoke Prior Boat Services** Stoke Wharf, near Bromsgrove, Worcs. (31401). Water, diesel, sewage disposal. Small slipway, permanent moorings and winter storage. Crane up to 1½ tons. Canal wares for sale.

PUBS

Tardebigge ¼ mile E of tunnel mouth, on A448. Food. Shop nearby.

Queen's Head Canalside, at bridge 48. Terrace overlooking canal; groceries next door.

Navigation Stoke Wharf, by the railway bridge.

DIGLIS BASIN

Enlargement from map page 101.

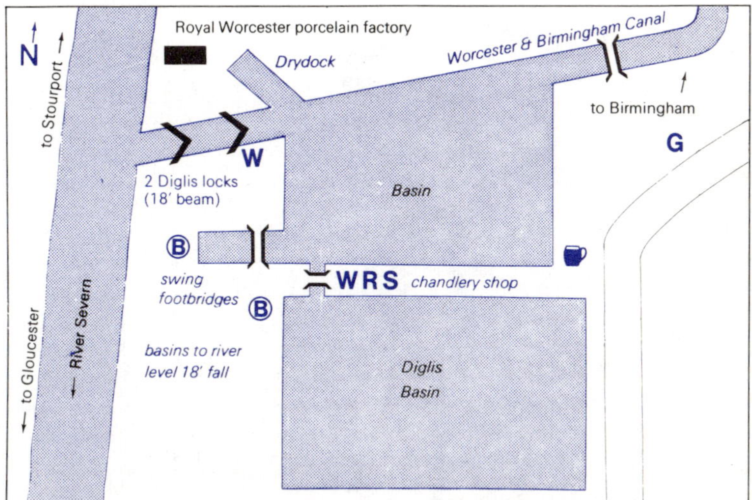

'Gongoozlers' at Tardebigge locks.

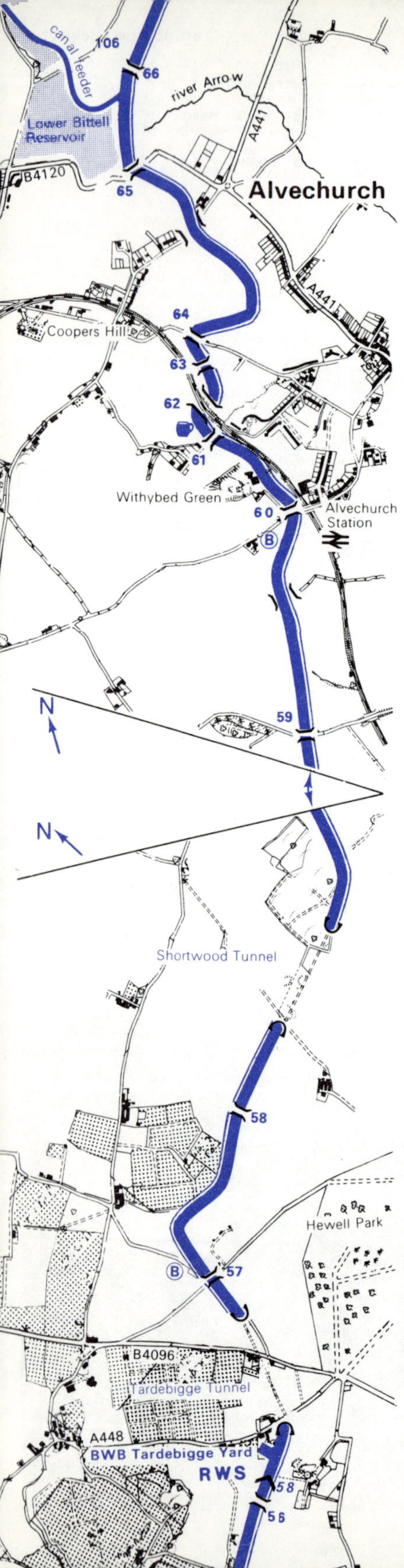

Alvechurch

$4\frac{3}{4}$ miles

This is a most delightful stretch of canal, that winds ceaselessly through the hilly Worcestershire countryside. The flat Severn valley seems very distant as the canal plunges first into Tardebigge and then Shortwood Tunnel. There is a boatyard between the tunnels. East of Shortwood Tunnel and the surrounding fruit plantations, the canal emerges high up on the side of a low wooded hill, overlooking the modest valley of the river Arrow. In the distance is the hum of traffic on the A441. The canal continues northward, winding steadily through this tranquil landscape until the small town of Alvechurch is reached. The town is set below the canal in a hollow, its church up on a hill; the canal winds tortuously along the steep hills round the outskirts, passing a boatyard, a station, an old brickworks and a charming canal pub. Then the canal turns abruptly north; ahead in the distance is the ridge of hills that is pierced by King's Norton Tunnel. Unexpectedly, an aqueduct carries the canal over a little lane that leads to Barnt Green and its station (a mile to the west). Past the aqueduct and through bridge 65, Lower Bittell Reservoir comes into view, beside and below the navigation. The canal crosses the valley on an embankment; at the north end of this is a very pretty red cottage which stands at the point where the feeder from Upper Bittell Reservoir enters the canal. With these on two sides and an overflow weir and the Lower Reservoir on the third, the house seems to be virtually surrounded by water.

Bittell Reservoirs
These 2 reservoirs were built by the canal company, the upper to feed the canal, the lower being a compensation to local mill owners for the loss of water resulting from construction of the canal. The reservoirs are nowadays popular among anglers and bird watchers; all sporting rights are let to Barnt Green Waters Ltd. (Honorary Treasurer: E. R. Greey, Peewit Cottage, Bittell Farm rd, Barnt Green, Worcs.)

Alvechurch
Worcs. PO, tel, stores, garage, bank, station (on branch line from Barnt Green to Redditch). A pleasant little town with some fine half-timbered houses, Alvechurch is situated at the bottom of a hollow and surrounded by folds of green hills. However through traffic on the A441 does not improve the place and there are plans for a southern bypass. The church stands alone on a hill; it is of Norman origin but was largely rebuilt by Mr Butterfield in 1861. There are some interesting monuments within.

Shortwood and Tardebigge Tunnels
608yds long and 568yds long respectively, these are 2 out of the 4 tunnels on the 14-mile summit level of the Worcester & Birmingham Canal. Neither contains a towpath, and until the turn of the century a company tug used to pull all boats through Tardebigge, Shortwood and the great King's Norton Tunnel. Boatmen will find Shortwood Tunnel extremely wet.

BOATYARDS & BWB

Ⓑ **Warwickshire Cruisers** Scarfield Wharf, Alvechurch, Worcs. (021 445 2909). At bridge 60. Marine engineering of all kinds; boat fitting out, repairs and sales. Engine installation and repairs. Water, diesel, petrol, gas, chandlery, dustbins and sewage disposal. Slipway, permanent moorings, winter storage. Groceries nearby, over canal bridge.

Ⓑ **Tardebigge Boat Company** The Old Wharf, Tardebigge, Bromsgrove, Worcs. (Bromsgrove 73898). Marine engineers; boat building. Lister diesel engines repaired. Converted narrowboats for hire.

PUBS

Crown Alvechurch. Canalside, at bridge 61. Pleasant country pub.

King's Norton Tunnel

4⅝ miles

Leaving the reservoirs, the canal curves through a slight cutting to Hopwood, where there is a pub, a boatyard and a busy main road crossing. North of here the canal enters a cutting that leads to King's Norton Tunnel. The ridge of hills that this tunnel penetrates serves as an important geographical boundary; for to the south of it is the rolling open countryside of rural Worcestershire, while north of the tunnel is Warwickshire, and the southernmost indications of Black Country industrial development. The built-up area is revealed as soon as the canal leaves the cutting at the north end of the tunnel: to the west is King's Norton, while all around are new houses and light industries. At bridge 71 (the best access point for the village and its shops) is the main mooring site of the local boat club; just past it is the old canal cottage at King's Norton Junction. Here the Stratford-on-Avon Canal enters at right angles—the strange mechanism of the celebrated King's Norton stop lock can be seen a few hundred yards along it. Boats heading for the south east should turn off down the Stratford Canal here *(see page 131)*. Meanwhile the Worcester & Birmingham Canal continues towards Birmingham, under a roving bridge at the junction, then under an inelegant steel bridge that connects a canalside factory with its car park. Another turnover bridge (73) is encountered: there is a telephone box near it and a further industrial stretch just beyond.

King's Norton
Worcs. PO, tel, stores, garage, bank, station. The village has done well to survive as a recognisable entity, for the suburbs of Birmingham have now extended all around it, and the amount of urban traffic passing through the village leaves one gasping. But it *is* still a village, and the small village green, the old grammar school buildings and the soaring spire of the church ensure that it will remain so. The church is set back a little from the green in an attractive churchyard, and is mainly of the 14thC, although 2 Norman windows can still be seen. The grammar school is even older—it was probably founded by King Edward III in 1344. An interesting puzzle is that the upper storey is apparently older than the ground floor. . . . The school declined during the last century and was closed in 1875. It was restored 20 years ago and is now an ancient monument.

King's Norton Tunnel
Otherwise known as Wast Hill or West Hill Tunnel, this 2726-yd bore is one of the longest in the country. It is usually difficult to see right through the tunnel, and there are plenty of drips from the roof in even the driest weather. A steam-powered—and later a diesel—tunnel tug service used to operate in the days of horse-drawn boats (there is no towpath). The old iron brackets and insulators that still line the roof were installed to carry telegraph lines through the tunnel. Grandiose bridges (nos. 69 and 70) span the cutting at either end.

Hopwood
Worcs. Tel, garage. More a name than a village, Hopwood is merely a small settlement. There are buildings near the canal which may be useful to canal travellers; a pub, a small boatyard, a petrol station. A fast main road bisects the area.

BOATYARDS & BWB

Ⓑ **Hopwood Craft** Birmingham road, Hopwood, near Alvechurch, Birmingham. (021 445 2595). Water, diesel, dustbins. Small slipway; light boat repairs. Outboard and inboard engine repairs. Boatbuilding and fitting out at their main yard at Horninglow Wharf, Burton upon Trent (on the Trent & Mersey Canal).

PUBS

Navigation Inn King's Norton, 100yds W of bridge 71.

Hopwood House Hopwood. Canalside, at bridge 67. Food usually available.

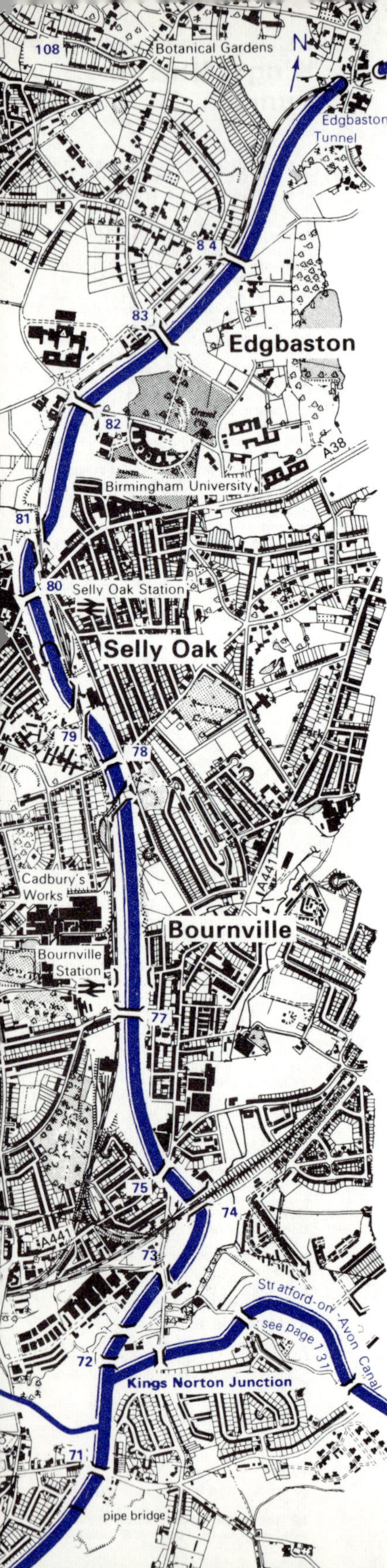

Edgbaston

$4\frac{3}{4}$ miles

Just north of King's Norton Junction, the canal enters a grimy industrial area, where chimneys belch, hidden machinery roars and steams, and the canal is left like an uninvited guest. Access is closed off at most of the bridges. Canal veterans will recognise this as typical of an approach to the great city of Birmingham and will doubtless resign themselves to 10 miles of this kind of scenery. But, thankfully, it does not last, for the canal seems to hold the industries at bay on one side, while a railway line (the main line from Worcester and the South West to Birmingham) draws alongside on its west flank. Canal and railway together drive through the middle of Cadbury's Bournville works, which is interesting rather than oppressive. Beyond it is the rather tired-looking Bournville station, followed by a cutting. Soon the railway vanishes briefly behind the buildings of Selly Oak; the canal goes through this suburb, but access is closed off at the road bridge (80). Between this and the next, skewed, railway bridge is the site of the junction with the Dudley Canal, but no trace remains here now of either the junction or the canal itself. North of here the canal and railway together shrug off industry and town, and head off on an embankment towards Birmingham in splendid isolation and attractive surroundings. Below on either side is the green spaciousness of residential Edgbaston, its botanical gardens and woods. A hospital is on the west side. The University of Birmingham is on the east side; among its many large buildings the most conspicuous is the Chamberlain campanile tower, which was erected in 1900. At one of the bridges near the University, 2 Roman forts used to stand; but most evidence of them was obliterated by the building of the canal and railway. Only a reconstructed part of the larger fort exists now. Past the University's moorings, canal and railway enter a cutting, in which their enjoyable seclusion from the neighbourhood is complete; the charming old bridges are high, with no access possible, while the cutting is steep, and always lined by overhanging foliage. It is a remarkable approach to Birmingham, and quite unrivalled by any other waterway. The railway is the canal's almost constant companion, dipping away here and there to reappear a short distance further on; but trains are not too frequent, and in a way their occasional appearance heightens the remoteness that attaches to this length of canal. At one stage the 2 routes pass through short tunnels side by side: the canal's tunnel is the northernmost of the 5 on this canal and the only one with a tow-path through it. It is a mere 105yds long.

Edgbaston
A desirable residential suburb of Birmingham, Edgbaston is bisected but unnoticed by the canal, and there is little contact between them. However, for those who contrive to be on the 'landward' side of Edgbaston, there are several things to be visited:

Cannon Hill Park Edgbaston, about $1\frac{1}{2}$ miles east of the University. Formal gardens including a Japanese Garden of Contemplation; tropical and sub-tropical plants adjacent. Birmingham Zoo is now housed here.

Cannon Hill Museum Pershore rd, Edgbaston. Designed primarily for children. Illustrated leisure time pursuits including bird-watching, bee-keeping, fishing and pets. Safari hut around which the sounds, sights and smells of the African bush are re-created. *Open weekdays 10.00-20.00 (in winter 10.00-17.00), Sun 14.00-17.00.*

Geological Department Museum The University, Edgbaston. Collection of palaeontology, stratigraphy, petrology, mineralogy and physical geology, including the Holcroft collection of fossils and the

Lapworth collection of grapholites. *Open daily 9.00-17.00 by arrangement.* (021 472 1301).

Botanical Gardens Edgbaston. Founded over 100 years ago. Alpine Garden, lily pond and a collection of tropical birds. *Open weekdays 9.00-dusk, Sun 10.30-dusk.*

Perrott's Folly Monument rd, Edgbaston. Seven-storey tower built in 1758 by John Perrott. One theory as to its origin is that Mr Perrott could, from its height, gaze on his late wife's grave 10 miles away. Since the late 1800's it has been used as an observatory.

The Dudley Canal
This canal used to join the Worcester & Birmingham at Selly Oak, thus providing a southern 'bypass' round Birmingham, but although interest in the west end is now reviving, the eastern end of the canal has been closed for many years, and will certainly remain so. The tremendously long (3795yd) Lappal Tunnel (now collapsed) emerged 2 miles from Selly Oak. This bore was more like a drain pipe than a navigable tunnel—it was only 7ft 9in wide, a few inches wider than the boats that used it, and headroom was limited to a scant 6ft 0in. Boats were assisted through by a pumping engine flushing water along the tunnel, but it must still have been a nightmarishly claustrophobic trip for the boatmen.

Bournville Garden Factory
The creation of the Cadbury family who moved their cocoa and chocolate manufacturing business south from the centre of Birmingham. The Bournville estate was begun in the late 1800's and is an interesting example of controlled suburban development. The old canal wharves can be clearly seen, but nowadays most of the ingredients travel by rail—the shunting engines around here are painted in the familiar Cadbury's livery.

Selly Manor and **Minworth Greaves**
Bournville. Two half-timbered Birmingham houses of the 13thC and early 14thC, re-erected in the 1920's and 30's in Bournville. They contain a collection of old furniture and domestic equipment. *Open Tue, Wed, Fri 13.00-17.00.* Enquiries to the Curator, 44 Mulberry rd, Bournville, Birmingham, B30 1TA. *The nearest point of access from the canal is at Bournville station: walk west to the Cadbury's entrance. There is a public right of way (Birdcage Walk) through the works: bear right at the fork, then turn right at the village green. The 2 houses are close by, on the left.*

The Worcester & Birmingham Canal's secluded course through Edgbaston.

Worcester & Birmingham
Birmingham & Fazeley

Birmingham

$4\frac{1}{4}$ miles

The Worcester & Birmingham canal now completes its delightful approach to Birmingham, and in only the last few hundred yards to Worcester Bar Basin (Gas Street Basin) does it assume the appearance of a typical Birmingham waterway. The railway disappears underneath in a tunnel to New Street station, while the canal suddenly makes a 90-degree turn left to the basin. The terminus of the Worcester & Birmingham Canal is the former stop lock; this is known as Worcester Bar, for originally there was a physical barrier here between the Worcester & Birmingham Canal and the much older Birmingham Canal: the latter refused to allow a junction, and for several years goods had to be transhipped at this point from one canal to the other. This absurd situation was remedied by an Act of Parliament in 1815, by which a stop lock was allowed to be inserted to connect the two canals. Nowadays the stop gates are kept open and one can pass straight through, on to the Birmingham Canal. This canal goes under a short tunnel (or a long bridge) with a church on top of it, to Farmers Bridge Junction. From here the Birmingham Canal aims off north west, towards the body of the Birmingham Canal Navigations network *(see page 116)*.

The route described below is the Birmingham & Fazeley Canal from Farmers Bridge to Salford Junction, where it links up with the canals described in book 1 of this series.

Turning north east off the main line of the Birmingham Canal, one arrives shortly at Cambrian Wharf – a startlingly attractive canal basin with a modern canal pub, BWB moorings and a generally smart and tidy appearance, all overlooked by 4 big blocks of flats. There are some well-painted locks too, for from this point the Farmers Bridge flight of 13 locks descends steeply into the heart of Birmingham. Many of the locks were built very close together and so the intervening pounds were expanded as much as possible in every direction. This results in one side of each lock becoming like a peninsula, flanked by water. The course of the flight is mainly dark and mysterious, for over the years the canal has become so frequently crossed by bridges, and even roofed over by vast buildings, that it becomes like a subterranean journey – and not always an easy one, working through locks where daylight is never abundant. The canal is however totally cut off from the town that jostles it so closely: all access places have been sealed off, and consequently towpath walkers are rarely seen. After passing the base of Birmingham's Post Office Tower (which is tall, cylindrical and spiky) and under the great arch of the now closed Snow Hill station, the canal levels out as the locks come to an end. But soon comes Aston Junction, marked by the old iron turnover bridge, on which is cast 'Horseley Iron Works Staffordshire 1828'. To the north east is the main line of the Birmingham & Fazeley, falling through the 11 Aston locks to Salford Junction. The scenery is typical Birmingham, and is not very prepossessing: the canal banks are lined by the back walls of industries, the old private canal arms and basins are piled off or choked with weeds. One of the bridges along here has girders that restrict the headroom to 7ft 9in; nearby is a canalside petrol station. At the second lock from the bottom is a lock keeper's cottage: its gate onto the road gives access to a small grocery, a petrol station and a pub. This is probably the only access point between Farmers Bridge and Salford Junction: a pity, since as on the rest of the Birmingham canals the towpath is in excellent condition. Towards Salford Junction itself, the buildings become fewer and lower. A half-sunken narrowboat along here provides a wonderful natural water-garden for many flowering weeds.

Salford Junction
At this unusual junction the Birmingham & Fazeley arrives from Birmingham, crossing the dirty river Tame on an aqueduct and bearing half right (east) towards Minworth and Fazeley. Meanwhile the short Birmingham & Warwick Junction Canal (the Saltley Cut) appears from the Grand Union Canal to the south, also crossing the river Tame on a low aqueduct. The straight canal entering the junction from the north-west is the Tame Valley Canal, part of the BCN system. However, this intersection of waterways is of little significance compared to what is going on overhead, for this is the site of the notorious Gravelly Hill interchange – the most elaborate urban road junction in Britain. So the whole area of Salford Junction is completely overpowered by the mass of concrete pillars and decking carrying the endless lines of unseen vehicles in all directions. The sky is truly filled with roads, while the little-known canals creep along underneath. The noise is a little disheartening, but it is a fascinating way to see this remarkably lavish piece of civil engineering.

The Digbeth Branch
This leaves the Birmingham & Fazeley main line at Aston Junction, and descends through 6 locks to Bordesley Basin, now disused, where it meets the former Warwick & Birmingham Canal, which became part of the Grand Union Canal when the G.U.C. Company was formed in 1929. There was a stop lock – called Warwick Bar – at the junction by Bordesley Basin. One of the lesser-known tunnels on the canal system is on the Digbeth branch – this is Ashted Tunnel, a short and gloomy tube in the middle of the Ashted locks. There is a narrow towpath through the tunnel, protected by railings and with a corrugated surface for the towing horses to get a good grip on.

Cambrian Wharf
The winner of a Civic Trust award in 1970, this development is a promising vision of what can be done to revive Birmingham's canals and create an exciting new townscape for the inhabitants. The construction of four new tower blocks of flats near Farmers Bridge (very close to the centre of Birmingham) provided the incentive for BWB to dredge out and restore the basin at the beginning of the former Newhall branch canal; for Birmingham City Council to make a pleasing canalside walk and restore two 18thC terraces of cottages in the nearby Kingston Row; and for a local brewery company to build an inspired new canal pub with an extra bar in a floating narrowboat. The overall effect is excellent.
Apart from Gas Street Basin, Cambrian Wharf is the best place – the only place – to leave a boat moored safe from the attentions of wandering vandals. There are permanent moorings available here: apply to the BWB Canal Shop in Kingston Row. *(See below).* For walkers, it is unfortunately impossible to go by canal from Cambrian Wharf to Gas Street Basin *(see below)*, since the towpath is sealed off at Cambrian Wharf. One must therefore go through the streets, and the route is as follows: Cambrian Wharf, Kingston Row, St Martin's Place, Cambridge Street, straight across Broad Street, down Gas Street and turn left through a small opening. This leads down onto the towpath at the Basin.

Worcester Bar Basin
Unofficially known as Gas Street Basin, it is an amazing contrast with Cambrian Wharf. For, whereas the latter is smart and new, Gas Street is old, comfortable, secluded and very crowded. It is full of delightful narrowboats in working trim but mostly out of work now. The basin is shaped like a flat triangle, with a causeway across the middle. Totally hemmed in by original canal warehouses of all kinds, Gas Street Basin has an unassailably authentic atmosphere and is a remarkable settlement to find surviving in the centre of such a contemporary city as Brum.

BOATYARDS & BWB

BWB Canal Shop & Information Centre 2 Kingston Row, Birmingham B1 2NU (021 236 4844). Chandlery, canal publications and wares for sale; BWB information about the hire of BWB cruisers from Nantwich and Hillmorton, and leaflets about canal cruising in general. Moorings and facilities (water, refuse & sewage disposal) at Cambrian Wharf administered from here. *Closed Sun.*

BOAT TRIPS

'The Brummagem Fly' – an early canal passenger service – has been recently restored in the shape of a 1935 narrowboat (formerly a BWB waterbus operating in London). The boat runs day trips for private parties and a public service (maximum 48 passengers) from Farmers Bridge, Birmingham. Trips last from 1 to 6 hours. Educational cruises a speciality. Food and drink on board can be arranged. All enquiries to 021 643 0525.

PUBS

Duke of Wellington near the second lock from Salford Junction, on the Birmingham & Fazeley Canal. The only pub easily accessible from the canal between Farmers Bridge and Salford.

Long Boat Cambrian Wharf, Birmingham. Canalside, near the top lock. Large new canal pub, whose 'garden furniture' includes an old canal crane and dock gates. Interior fittings include an original Bolinder engine and butty boat rudders. One of the bars is in a floating narrowboat. Food usually available at bar.

The Opposite Lock Gas Street Basin. A canalside pub for motor-racing types. Others (like boaters) may be admitted if they are presentable. Wining, dining and dancing. *Mon-Sat until 2.00.* Enquiries to 021 643 2573.

Birmingham & Fazeley Canal. The towing path rises here over one of the many derelict private basins leading off the canal in Birmingham: the 'rib' formation of the brickwork was designed to assist the towing horses get a firm grip on the slope.

Maximum dimensions

Length: 71′ 6″
Beam: 7′
Headroom: 6′ 6″

Mileages

Birmingham Canal new main line
BIRMINGHAM Gas Street to SMETHWICK JUNCTION (old main line): $2\frac{7}{8}$
BROMFORD JUNCTION: $4\frac{7}{8}$
PUDDING GREEN JUNCTION (Wednesbury Old Canal): $5\frac{5}{8}$
TIPTON FACTORY JUNCTION (old main line): $8\frac{3}{8}$
DEEPFIELDS JUNCTION (Wednesbury Oak Loop): 10
Bradley Workshops: $2\frac{1}{4}$
HORSELEY FIELDS JUNCTION (Wyrley & Essington Canal): 13
Wolverhampton top lock: $13\frac{1}{2}$
ALDERSLEY JUNCTION (Staffordshire & Worcestershire Canal): $15\frac{1}{8}$

Total 24 locks (3 up, 21 down)

Birmingham Canal old main line
SMETHWICK JUNCTION to
Junction with Engine Branch: $\frac{1}{2}$
SPON LANE JUNCTION: $1\frac{1}{2}$
OLDBURY JUNCTION (Titford Canal): $2\frac{1}{2}$
BRADES HALL JUNCTION (Gower Branch): $3\frac{1}{2}$
Aqueduct over Netherton Tunnel branch: $4\frac{3}{8}$
TIPTON JUNCTION (Dudley Canal): $5\frac{1}{2}$
FACTORY JUNCTION (new main line): 6

Total 3 locks

Dudley Canal line no. 1
TIPTON JUNCTION to
Dudley Tunnel (north end): $\frac{3}{8}$
PARK HEAD JUNCTION: $2\frac{3}{8}$
Black Delph bottom lock (Stourbridge Canal): $4\frac{1}{2}$

Total 12 locks

Dudley Canal line no. 2
PARK HEAD JUNCTION to
WINDMILL END JUNCTION: $2\frac{5}{8}$
COOMBESWOOD BASIN: $5\frac{1}{2}$

No locks

Netherton Tunnel Branch
WINDMILL END JUNCTION to DUDLEY PORT JUNCTION: $2\frac{7}{8}$

No locks

Wednesbury Old Canal
PUDDING GREEN JUNCTION to RYDER'S GREEN JUNCTION: $\frac{5}{8}$

No locks

Walsall Canal
RYDER'S GREEN JUNCTION to
Ryder's Green bottom lock: $\frac{3}{4}$
TAME VALLEY JUNCTION: $1\frac{3}{8}$
Junction with Anson Branch (for Bentley Canal): $5\frac{1}{8}$
WALSALL JUNCTION: $6\frac{7}{8}$

Total 8 locks

Walsall Branch Canal
WALSALL JUNCTION to BIRCHILLS JUNCTION (Wyrley & Essington Canal): $\frac{7}{8}$

Total 8 locks

Wyrley & Essington Canal
HORSELEY FIELDS JUNCTION to WEDNESFIELD JUNCTION (Bentley Canal): $1\frac{1}{4}$
SNEYD JUNCTION: $6\frac{1}{4}$
BIRCHILLS JUNCTION (Walsall Branch Canal): 8
PELSALL JUNCTION (Cannock Extension): $12\frac{7}{8}$
Norton Canes Docks: 2
CATSHILL JUNCTION: $15\frac{3}{8}$
OGLEY JUNCTION (Anglesey Branch): $16\frac{3}{8}$
Anglesey Basin and Chasewater: $1\frac{1}{2}$

No locks

Daw End Branch
CATSHILL JUNCTION to LONGWOOD JUNCTION (Rushall top lock): $5\frac{1}{4}$

No locks

Rushall Canal
LONGWOOD JUNCTION to RUSHALL JUNCTION: $2\frac{3}{4}$

Total 9 locks

Tame Valley Canal
TAME VALLEY JUNCTION to RUSHALL JUNCTION: $3\frac{1}{2}$
Perry Barr top lock: $5\frac{1}{2}$
SALFORD JUNCTION: $8\frac{1}{2}$

Total 13 locks

The Birmingham Canal Company was authorised in 1768 to build a canal from Aldersley on the Staffordshire & Worcestershire Canal to Birmingham. With James Brindley as engineer the work proceeded very fast. The first section, from Birmingham to the Wednesbury collieries, was opened in November 1769, and the whole 22½-mile route was completed in 1772. It was a winding, contour canal, with twelve locks taking it over Smethwick, and another twenty (later 21) taking it down through Wolverhampton to Aldersley Junction. As the route of the canal was through an area of mineral wealth and developing industry, its success was immediate. Pressure of traffic caused the summit level at Smethwick to be lowered in the 1790s (thus cutting out 6 locks–3 on either side of the summit), and during the same period branches began to reach out towards Walsall via the Ryder's Green locks, and towards Fazeley. Out of this very profitable and ambitious first main line, there grew the Birmingham Canal Navigations, more commonly abbreviated to BCN. After a long dispute about the building of the canal from Birmingham to Fazeley, the Birmingham Company bought up the embryonic Birmingham & Fazeley Canal Company; in 1794 the cumbersome title created by this merger, 'The Birmingham and Birmingham & Fazeley Canal Company', was changed to the simpler BCN. The battle for the right to build the Fazeley line was long and hard, and was fought with considerable intrigue and bitterness, a pattern of behaviour that tended to surround all the activities of the Birmingham Company. Being first in the field, it generally behaved in a high-handed manner, holding the whip hand over rivals when extensions and developments were proposed. Generally it exacted high compensatory tolls, and exercised strict controls over water rights–habits that pleased the shareholders but infuriated competitors.

The BCN network that exists today developed from three rival companies each seeking to capture traffic from the others. This intense competition resulted in a very intricate network, which was thus able to cater for all the material and distribution needs of the developing industries in the area. The web of lines forming the BCN became the veins of the Black Country, carrying the life blood of its commerce and wealth.

Apart from the Birmingham Company, there were two other companies instrumental in the creation of the BCN network. Over the other side of the Rowley Hills, the Dudley and Stourbridge Companies had set up a rival route to the Staffs & Worcs, to the annoyance of the Birmingham Company. Thomas Dadford, who had worked with Brindley earlier, was appointed engineer, and in three years (1776–9) it was completed as planned to a point just below the present Blower's Green Lock. Immediately there were plans to extend it underground to link with Lord Ward's private canal, and with the Birmingham at Tipton. After several setbacks this extension was completed between 1785 and 1792, including the long Dudley tunnel. Then once again the directness of the Dudley company prompted it to undertake a further extension to link with the recently authorised Worcester & Birmingham Canal at Selly Oak, a means of avoiding the severe compensation tolls exacted by the Birmingham Company for the junction at Tipton. This new line, eleven miles long, was opened in 1798. It included two tunnels, that at Lappal being the fifth longest in Britain. Cut through rock strata with great difficulty, this tunnel suffered continuously from subsidence and roof falls, and had to undergo frequent closure for repairs. The financial strain of this last extension nearly crippled the Dudley Company, and it only just managed to survive until 1846, when it was absorbed by the BCN. Lappal tunnel was finally closed in 1917.

Up in the north, the Wyrley & Essington Company joined the fray, completing a line from Wolverhampton to Wyrley in 1795 under the direction of William Pitt. This company also grew quickly, extending the canal first to Brownhills, and then to join the Coventry Canal at Huddlesford via the Ogley flight of thirty locks. Several branches were added to serve the rich coalfields around Cannock and Brownhills, which were destined to serve the BCN and West Midlands well when the Black Country pits began to decline. Indeed the meandering line of the Wyrley & Essington saw some of the last commercial traffic on the whole BCN network. The Birmingham Company had spread northwards to Walsall, but because of ill-feeling and rivalry, the logical link with the Wyrley & Essington line was not made until 1840, when the Walsall Branch Canal was built.

Traffic continued to increase, and with it the wealth of the BCN. The pressures of trade made the main line at Smethwick very congested, and brought grave problems of water supply. Steam pumping engines were installed in several places to recirculate the water, and the company appointed Thomas Telford to shorten Brindley's old main line.

Between 1825 and 1838 he engineered a new main line between Deepfields and Birmingham, using massive cuttings and embankments to maintain a continuous level. These improvements not only increased the amount of available waterway (the old line remaining in use), but also shortened the route from Birmingham to Wolverhampton by seven miles.

Serious congestion had also arisen at Farmer's Bridge locks, which could not keep up with the traffic although they operated 24 hours a day and 7 days a week. Land was not available for a duplicate flight in the immediate area, and so an earlier plan to build a canal following the valley of the river Tame was revived. However, it was not until railway control and amalgamation with the Wyrley & Essington had come in 1840 that the necessary impetus came to promote the Tame Valley Canal, and the whole series of extensions and improvements to the network that accompanied it. These developments led to the building of a relief for the narrow Dudley Tunnel: the Netherton tunnel, cut on a parallel course, included a towpath on each side and gas lighting throughout. The last addition to the network was the Cannock Extension Canal to Hednesford Basin, with its link to the Staffs & Worcs Canal via Churchbridge locks.

Railway control of the BCN meant an expansion of the use of the system, and a large number of interchange basins were built to promote outside trade by means of rail traffic. This was of course quite contrary to the usual effect of railway competition upon canals. Trade continued to grow in relation to industrial development, and by the end of the 19thC it was topping $8\frac{1}{2}$ million tons per annum. A large proportion of this trade was local, being dependent upon the needs and output of Black Country industry. After the turn of the century this reliance on local trade started the gradual decline of the system as deposits of raw materials became exhausted. Factories bought from further afield, and developed along the railways and roads away from the canals. Yet as late as 1950 there was over a million tons of trade, and the system continued in operation until the abrupt end of the coal trade in 1966, a pattern quite different from canals as a whole. Even today there is a small amount of commercial traffic, some of it still horse-drawn—a dramatic contrast to the roaring traffic on the new Birmingham motorways.

As trade declined, so parts of the system fell out of use and were abandoned. At its heyday in 1865, the BCN comprised over 160 miles of canal. Today just over 100 miles survive, and of these some 65 are without a guaranteed future, being classed as 'Remainder Waterways'. However, all the surviving canals of the BCN are of great interest; although not ideal for leisure cruising, they represent a most vivid example of living history, one of the most important monuments to the industrial revolution.

The report of the official working party of 1970 indicates the value of the surviving BCN network for amenity and water supply purposes, and, at the same time, points to the enormous cost of eliminating the 'Remainder' sections; as always, restoration is cheaper than removal. The recent work on the reopening of the Dudley Tunnel line and Park Head locks shows the increasing local council interest in canals as a whole, and their concern for urban recreational facilities. If this pattern is to continue, then the future of this unique network would seem secure. Birmingham's canals are in places derelict and dirty but nearly all bring a vein of the countryside right into the heart of England's second largest city. The hub of the network is Gas Street Basin and here many varieties of fern grow on walls, posts, etc. Two of the commonest are polypody and hartstongue. From July until September, the beautiful pink flowers of the rosebay willow herb can be seen wherever there is a piece of waste ground. Fireweed, as it is called, was once very rare in England but since bomb-damage and demolition created many open spaces, it has become one of our most common and beautiful wild flowers. It is called 'fireweed' because it is a 'first coloniser', often appearing very soon after a fire has cleared an area, especially in woodland. This flower is much enjoyed by the caterpillars of the elephant hawk moth.

Also common is the yellow flowered ragwort which can be seen growing by the canal on walls and even in neglected gutters. Ragwort contains a poison which destroys the liver of any animal that eats it and has to be eradicated on all land on which livestock is kept. But it lends a touch of sunshine gold to many urban canals.

The bird life of Birmingham is surprisingly rich. If you are lucky you may see, beside blackbirds, thrushes, tits and wagtails, the black redstart with is grey head and back, black chest and bright rusty red tail feathers. Black redstarts are nowhere common in Britain but they have actually nested in the centre of Birmingham only a few hundred yards from Gas Street Basin itself. Kestrels are quite common in the city; they nest on the larger buildings and feed on small mammals and birds. Most evenings, as dusk begins to fall, flocks of starlings arrive in the city, commuting to their favourite roosting sites in the warmth from their feeding grounds in the surrounding countryside. Before finally settling down for the night, they perform one of the most spectacular flying displays ever seen, yet most human commuters scurry past oblivious of this beautiful aerial ballet over their heads. Up to 50,000 starlings roost in the city centre.

Moorhens nest quite close to the city and in the spring their untidy nests can be seen along the canal banks, always in imminent danger of being flooded by the wash from passing boats. Moorhens are very shy and often hide underwater with only their bills showing when danger threatens. Kingfishers are not uncommon on Birmingham canals and a pair has nested recently in the Edgbaston area. They feed on sticklebacks, which abound in the BCN and seem to be among the few fish which can survive the industrial pollution which periodically sweeps down these waterways. Although parts of the BCN suffer from industrial pollution, the fishing is very good in places. This is useful in an area of poor fishing facilities.

Sticklebacks themselves feed on the great variety of invertebrate life which lives in the seemingly dead waters of the canal. Wherever there is a patch of weed or reeds you can find dragonfly nymphs, water beetles, freshwater shrimps, leeches, frogs, newts and tadpoles.

Many small rodents make their homes along the banks. The most common are water voles, vegetarians feeding on reeds and grass; you can often see their tracks in the mud, and look out for their latrines since they are tidy animals and leave their droppings in one place. Their burrow systems are quite large, with at least two entrances—one being underwater for a quick exit in case of danger. The brown rat is also common and the water vole is often mistaken for it. Rats swim readily but they are not so skilful as water voles and always come out looking bedraggled since their fur is not waterproof. Voles and rats cause major problems through burrowing into the canal banks. This often leads to serious canal leaks.

Few people realise that Birmingham has a large fox population and if you are moored in the city at night, in early springtime you may hear the eerie wailing call of the vixen and the sharp yapping bark of the dog fox as they advertise their willingness to mate. Foxes are useful scavengers and help to keep down the rat population. The best place to see and hear them is on the Worcester & Birmingham canal as it runs through Edgbaston, and a good time is early in the morning as they return to their earths after the night's hunting. While you are listening for foxes, you will almost certainly hear the call of the Tawny owl proclaiming his territorial rights. Tawny owls are common in Birmingham, nesting in the many trees and old buildings that are to be found in the city.

Willows are interesting trees; apart from providing wood for cricket bats and osiers for basketmaking, their bark also contains salicin, a salt of salicilic acid. This is the major constituent of aspirin. On the canal, willows provide a refuge for many small birds, and food for many tiny animals, for example the larvae of the puss-moth. This beautiful bright green caterpillar with its velvety-brown back is well camouflaged. When disturbed, however, it extends two large bright red tails which it lashes about fiercely, rearing the front part of its body into the air and presenting its bright pink face to a potential enemy. If you are not deterred by this display, the caterpillar may as a last resort squirt formic acid at you.

These are just a few of the many fascinating plants and animals which can be found on the canals of Birmingham. Any enquiries concerning their natural history may be addressed to the Department of Natural History, the City Museum and Art Gallery, Birmingham, B33DH.

A canal 'flyover' on the BCN.

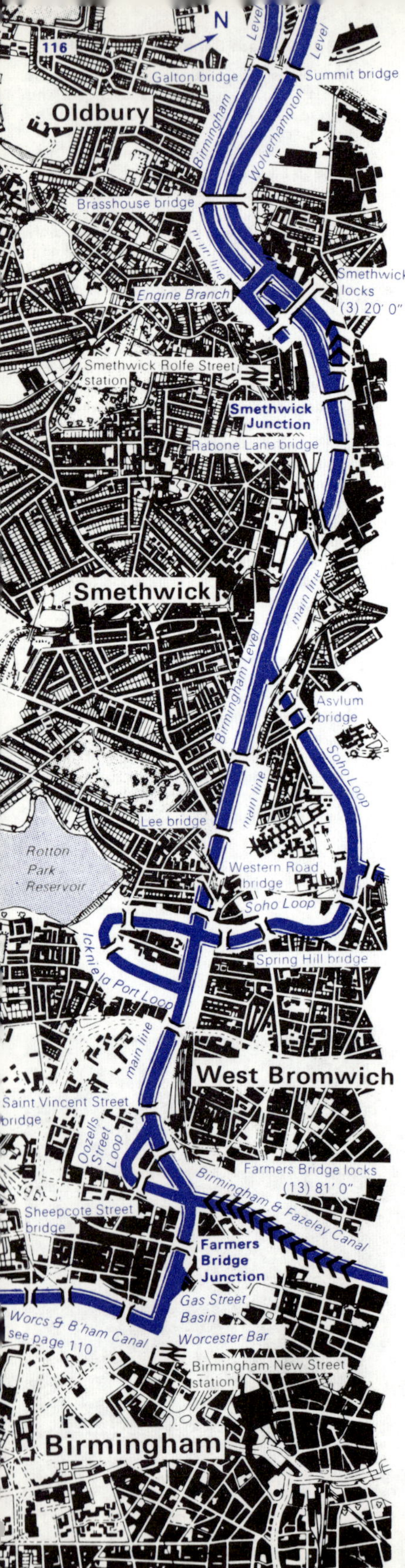

116 Birmingham Canal Main Line

Smethwick

$3\frac{1}{2}$ miles

The Worcester & Birmingham Canal terminates in Worcester Bar Basin (Gas Street), at the famous Worcester Bar (for details of the basin see page 111). The main line of the Birmingham Canal Navigations (BCN) leaves the basin, passing immediately under Broad Street bridge, which has been extended so many times that it is now virtually a tunnel; the towpath continues beside the canal. Walkers should remember that Gas Street is the only towpath access point in Birmingham itself that is not kept locked, and is the last one on the main line for several miles. Even the new development at Cambrian Wharf has no direct access to the towpath. North of Broad Street is Farmer's Bridge Junction, a canal crossroads. Here the Birmingham & Fazeley Canal swings away to the east, passing immediately Cambrian Wharf and the Longboat pub, and then starting the 13 lock descent to Aston Junction (see page 110). The main line turns west at Farmer's Bridge, while the short Oozells Street loop goes to the south, quickly disappearing behind old warehouses. This loop, and the others further along, are surviving parts of Brindley's original contour canal, now known as the Birmingham Canal Old Main Line, which pursued a rather meandering course. The delays caused by this prompted the Birmingham Canal Company to commission Telford to build a straighter line, the Birmingham Canal New Main Line. This was constructed between 1823 and 1838, and when completed reduced Brindley's old $22\frac{1}{2}$-mile canal to 15 miles. The Oozells Street loop reappears from the south, and then, after two bridges, the Icknield Port loop leaves to the south. This loop acts as a feeder from Rotton Park reservoir; adjoining it are the BWB maintenance yard and area offices. Originally built to supply water to the Birmingham Canal, the reservoir is now used for water sports and recreation as well. The loop rejoins after $\frac{1}{4}$ mile at another canal crossroads—the Winson Green loop, which leaves the main line opposite the Icknield Port loop. This last loop is the longest of the three, running in a gentle arc for over a mile before rejoining the main line again. It is also the only loop to have a towpath throughout its length. At its eastern end are Hockley Basins, formerly railway-owned. At the point where the loop rejoins, there is an island in the middle of the canal, the site of one of the many toll offices that existed throughout the system. The main line continues towards Smethwick, still flanked on both banks by factories and warehouses, none of which pays any attention to the canal. A railway crosses to the west, the busy electrified main line that accompanies the canal all the way to Wolverhampton, and then a gentle curve leads to Smethwick Junction. Here there is a choice of routes; Brindley's old main line swings to the right, while Telford's new main line continues straight ahead. The two routes run side by side, but the old line climbs to a higher level via the three Smethwick locks. Here there are two flights of locks side by side, the extra flight built by Smeaton to overcome the traffic hold up. Brindley's original flight is now derelict, and so boats should use the southern or left-hand locks—Smeaton's flight. Beyond the junction, Telford's new line enters a steep-sided cutting. This 40ft deep cutting enabled Telford to avoid the changes in level of the old line, and thus speed the flow of traffic. The two routes continue their parallel courses, the one overlooking the other, until the lower line passes under the Telford aqueduct. This elegant single span cast-iron structure carries the Engine Arm, a short feeder canal that leaves the old line, crosses the new line, and then turns back to the south for a short distance. This arm is named after the first Boulton & Watt steam pumping engine to be bought by the Birmingham Canal Company. This continued to feed the old summit level for 120 years. It was then moved to

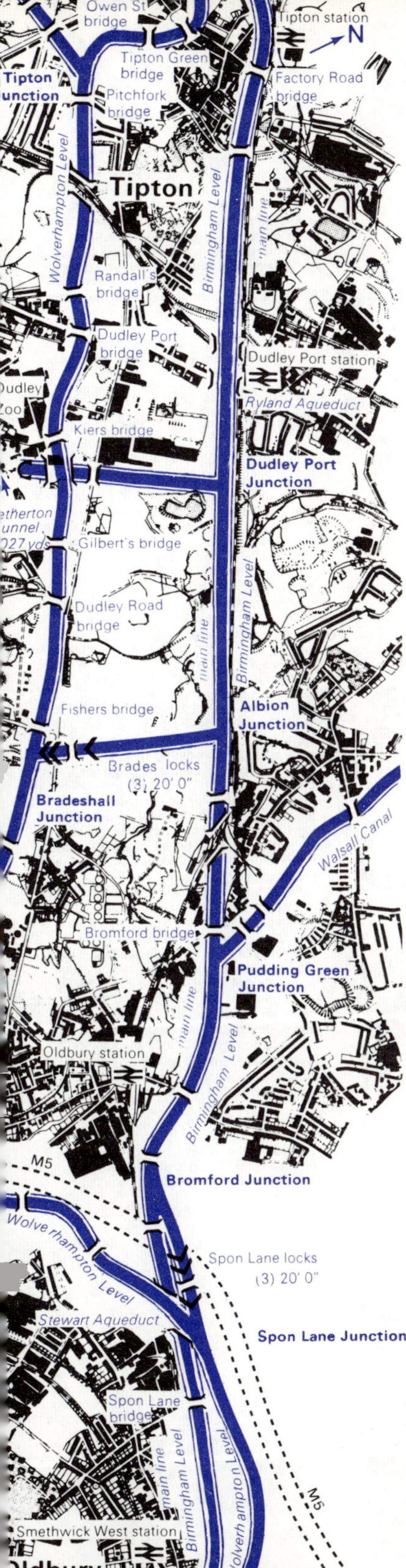

Birmingham Canal Main Line

Ocker Hill to pump water from disused mine workings until the 1950s when it was finally retired. (It is now in the Birmingham Museum of Science & Industry.) It is easy to walk from the new line up to the old at this point, although there is no access to the surrounding area from the canals. This isolation has meant that the canals are an oasis among the drab industrial surroundings. The sides of the cutting are richly covered with wild flowers and blackberry bushes, and the seclusion of the whole area has turned it into an unofficial nature reserve. The two canals continue through this natural wilderness to Telford's Galton bridge, which carries Roebuck Lane across the cutting in one magnificent 150ft cast-iron span. This famous bridge is to be closed to traffic and preserved as an ancient monument.

Dudley

4½ miles

The old and the new Birmingham Canal lines continue their parallel course, and soon the pleasant semi-rural isolation of the cutting ends, to be replaced by a complex meeting of three types of transport system. The M5 motorway swings in from the east, carried high above the canal on slender concrete pillars; the railway stays close beside Telford's new line; and the canals enter a series of junctions that seem to anticipate modern motorway practice. The new line leaves the cutting, and continues in a straight line through industrial surroundings. It passes under Stewart Aqueduct, and then reaches Bromford Junction. Here a canal 'sliproad' links the old and the new lines via the three Spon Lane locks, joining the new at an angle from the east. The old line swings south west, parallel to the M5 crossing the new line on Stewart Aqueduct. Thus canal crosses canal on a flyover. Spon Lane locks, the linking 'sliproad', survive unchanged from Brindley's day, and are among the oldest in the country. The old and the new lines now follow separate courses.
The old line continues below the motorway to Oldbury Locks Junction. Here the remains of the short Titford Canal climb away to the south via the six Oldbury locks; this canal serves as a feeder from Rotton Park reservoir. After the junction the old line swings round to the north west, and continues on a parallel course to the new line once again, but a mile to the south. At Bradeshall Junction the Gower branch links the two lines, descending to the lower level of the new line through three locks. At Tividale the old line crosses the Netherton Tunnel branch on an aqueduct, another canal flyover, and then continues to Tipton Junction. To the south west is the branch leading to Dudley Tunnel, reopened in 1973 after joint restoration by volunteers, BWB and Dudley Corporation. This branch connects with the Dudley Canal, the Stourbridge Canal, and thus with the Staffs & Worcs. The old line turns north at the junction, turning towards the new line which it rejoins at Factory Junction.
After Bromford the new line continues its straight, predictable and unexciting course towards Wolverhampton. At Pudding Green Junction the main line goes straight on; the right fork leads to Wednesbury, the Walsall Canal route to the Wyrley & Essington Canal, and the Tame Valley Canal to Rushall, and Salford Junction. Beyond Pudding Green the railway crosses to the north bank, staying close beside the canal. At Albion Junction the Gower branch turns south to join the old link at Bradeshall. At Dudley Port Junction the Netherton Tunnel branch joins the main line, having passed under the old line at Tividale. The tunnel mouth can be seen from the junction. The Netherton branch goes through the tunnel to Windmill End Junction; from here boats can either turn south down the old Dudley Canal to Coombeswood Basin, or west

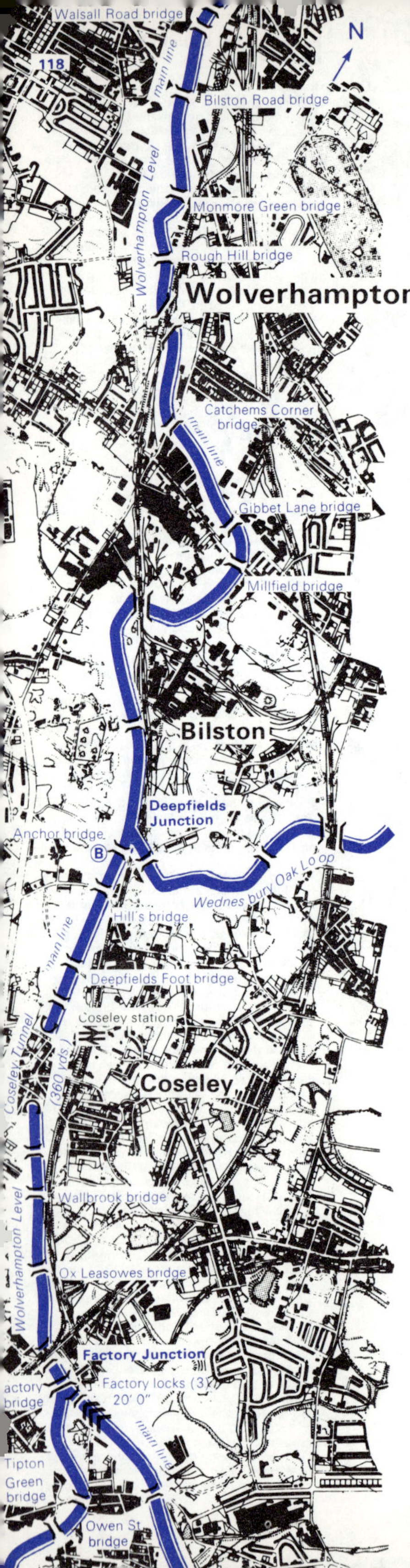

towards the Stourbridge Canal, and thus to the Staffs & Worcs. North of Dudley Port the new line crosses a main road on the Ryland Aqueduct; this structure was completely rebuilt in 1968, because the narrowness of the old aqueduct was hindering a road rebuilding scheme. One of the cast-iron name plates from the original aqueduct is preserved at the Waterways Museum, Stoke Bruerne. There is a convenient access point for walkers at the aqueduct, and all services for boats are close at hand. Continuing its elevated course the new line reaches Tipton, where access is again easy. Tipton station is right by the canal, and it is possible to walk from platform to towpath. Leaving the station, Factory bottom lock comes into view, the first of three which carry the new main line up to join the old Wolverhampton level.

Bilston

$4\frac{3}{4}$ miles

The new line climbs the three Factory locks, and then immediately reaches Factory Junction, where the old line comes in from the south. The long parallel course of the two canals is over. Uninviting and drab surroundings flank the canal as it approaches Coseley. Here there is a short tunnel, another sign of Telford's new route, for the old canal used to wind round Coseley Hill. The tunnel is of wide bore, with a towpath on each side; the style is the same as the much longer Netherton tunnel, opened in 1858, and indeed Coseley was cut to the same specification, as a trial run. Beyond the tunnel, the canal continues straight to Deepfields Junction. Here the old canal swings away to the east in what was once a long, meandering loop (the Wednesbury Oak loop) passing through the heart of the Black Country. The southern half of the loop has been filled in, leaving two miles of twisting canal which end at Bradley repair yard, a BWB maintenance depot where among other things, lock gates are made. Just before Deepfields there is a boatyard, with groups of narrow boats belonging to Alfred Matty, one of the few companies still operating boats commercially. There is easy access here, and from here to Wolverhampton access to and from the towpath at locks and bridges is much simpler. North of the junction the canal swings in a curve round Bilston steel works, whose furnaces and cranes fill the sky. Passing the works at night can be an exciting experience. The canal now continues its wandering course towards Horseley Fields Junction, its twisting and turning revealing that Brindley's original line was not altered by Telford north of Deepfields.

Navigational Note

Between Bromford Junction and Factory Junction the new main line is crossed by a series of two-arched bridges. Boats going towards Wolverhampton should pass through the northern, or *right-hand* arches. The other ones are shallow, and sometimes impassable. Likewise boats passing in the same direction should keep to the *right* of toll office islands.

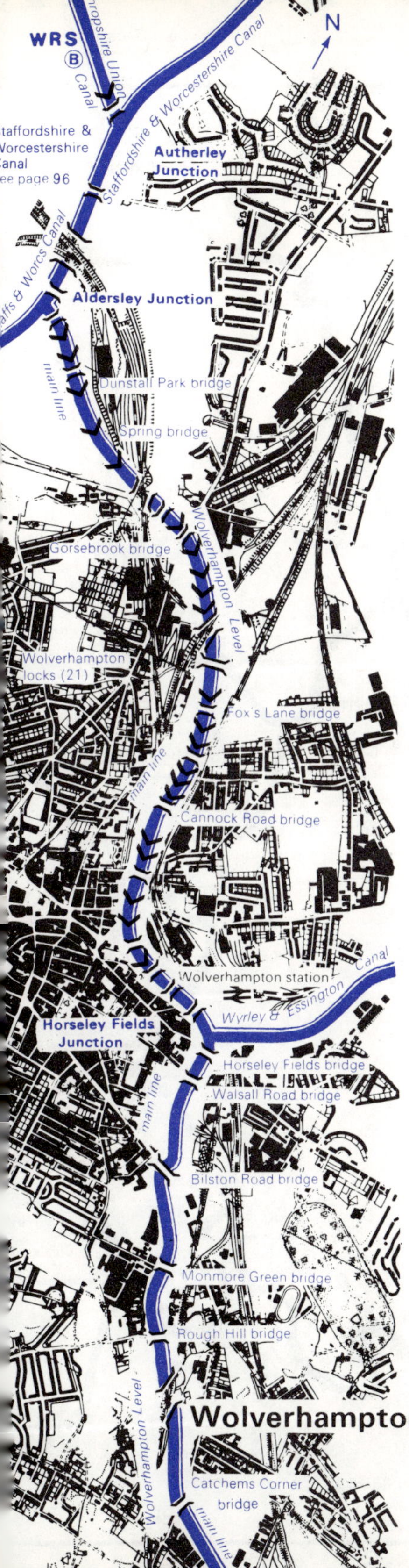

Wolverhampton

4 miles

The canal continues its winding course through the heart of industrial Wolverhampton. Factories surround the canal, shutting it off from the rest of the town, but access is not difficult. Just north of Bilston Road bridge there is a railway-canal interchange basin, still intact, the sidings running beside the covered wharf. It is a good reminder of the busy traffic that filled the BCN system until comparatively recently. At Horseley Fields Junction, set in the middle of Wolverhampton, the main line goes straight on. To the east, the Wyrley & Essington Canal starts its meandering contour course towards Brownhills. Originally this canal continued through Lichfield to join the Coventry Canal at Huddlesford, but now it has largely vanished east of Ogley Junction. However it is still possible to complete a wide circle via the Rushall and Tame Valley Canals. Soon after the junction the canal reaches the top lock of the Wolverhampton flight. From here 21 locks carry the canal down to join the Staffordshire & Worcestershire Canal at Aldersley Junction. The locks take the canal from the heaviest industry to quiet open countryside, a sharp contrast with the miles of industrial canal that lie to the south. Half the locks are flanked by industry, and railways criss-cross over and around the canal, but gradually this background yields until the last three locks which are virtually rural. Access is possible at several places in the flight, but the most convenient services are near the Cannock Road bridge and the Stafford Road bridge. Beyond the bottom lock there is a welcome old-fashioned brick arched bridge, and then the Birmingham Canal main line ends at Aldersley junction. (For the Staffordshire & Worcestershire Canal, see page 96.)

The BCN network

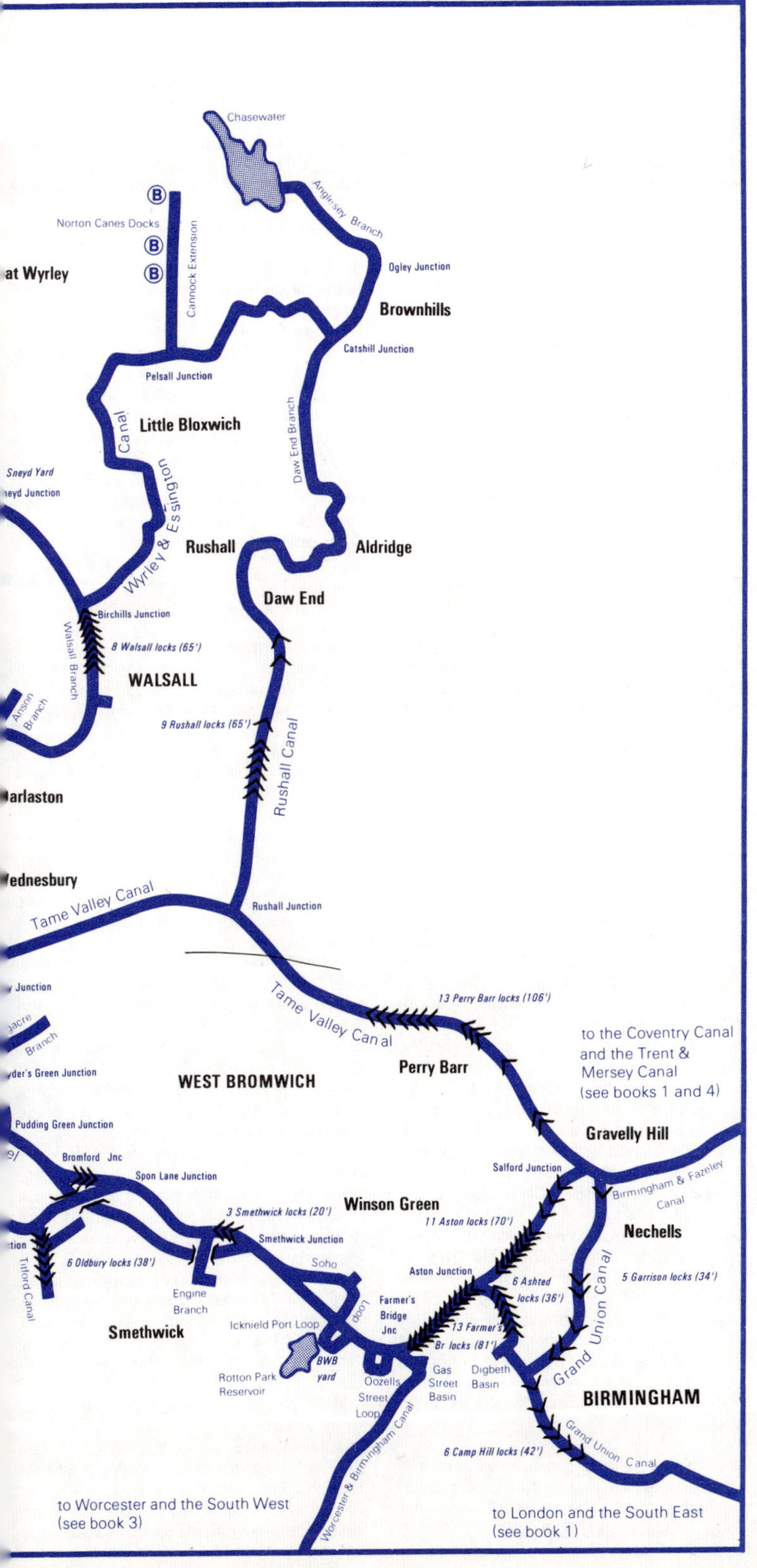
Chasewater
Anglesey Branch
Norton Canes Docks
Cannock Extension
Ogley Junction
Brownhills
Catshill Junction
Pelsall Junction
Little Bloxwich
Canal
Daw End Branch
Sneyd Yard
Wyrley & Essington
Rushall
Aldridge
Daw End
Birchills Junction
8 Walsall locks (65')
Walsall Branch
WALSALL
Anson Branch
9 Rushall locks (65')
Rushall Canal
Tame Valley Canal
Rushall Junction
Tame Valley Canal
13 Perry Barr locks (106')
Perry Barr
to the Coventry Canal and the Trent & Mersey Canal (see books 1 and 4)
WEST BROMWICH
Pudding Green Junction
Gravelly Hill
Bromford Jnc
Spon Lane Junction
Salford Junction
Birmingham & Fazeley Canal
Winson Green
3 Smethwick locks (20')
11 Aston locks (70')
Nechells
Smethwick Junction
Soho
Titford Canal
6 Oldbury locks (38')
Aston Junction
6 Ashted locks (36')
5 Garrison locks (34')
Engine Branch
Farmer's Bridge Jnc
Loop
Grand Union Canal
Smethwick
Icknield Port Loop
13 Farmer's Br locks (81')
BWB yard
Rotton Park Reservoir
Oozells Street Loop
Gas Street Basin
Digbeth Basin
BIRMINGHAM
Worcester & Birmingham Canal
Grand Union Canal
6 Camp Hill locks (42')
to Worcester and the South West (see book 3)
to London and the South East (see book 1)

Waterways of the BCN

Of the 100 odd miles of BCN waterway that are still open to navigation today, only six actually fall within the area of the City of Birmingham. The remainder of the surviving canals interlace nine local authority areas, and play a correspondingly vital role in industrial water supply and land drainage. Serving as arteries through the Black Country, the canals have an enormous recreational potential, although there is little commercial traffic using the network today. With the exception of the Birmingham main line, the Stourbridge and Birmingham & Fazeley lines, and Netherton tunnel, all the BCN is classed as 'remainder waterway'. In the report of the Birmingham Canal Navigations Working Party, published in 1970, it was recommended that much of the surviving network be promoted to 'cruising waterway' status, to ensure that the canals play their part in the developing leisure amenities for Birmingham and the Midlands generally.

Cruising the BCN is an unusual, and often dramatic, experience; the canals provide a continuous panorama of the history of the Black Country, and of the Industrial Revolution. There are frequent reminders of the importance of the canals in the development of the area, the toll islands isolated in the middle of the canal, the railway interchange basins, the factory arms and basins, often revealed today only by the arch of the towpath bridge buried in the brickwork, the loading bays of the factories which at one time were totally dependent upon the canals, the engine houses (the best example near the south portal of Netherton tunnel), and the disused and abandoned arms and branches, some closed so long ago that their course is now difficult to trace. Although industry has turned its back on the BCN, the canals are far from derelict or depressing. Scattered among the older buildings are new developments, many of which accept the canal as a natural part of the landscape. Cambrian Wharf is the obvious example, but in a different way the motorway interchange (commonly known as 'Spaghetti Junction'), built in the air above Salford Junction and the Tame Valley Canal, and the Post Office tower that dominates Farmer's Bridge locks are equally impressive. Not all the BCN is built up and industrial. For most of its length the Wyrley & Essington is a rural canal, following a winding contour line. The Rushall Canal and the Daw End Branch are also largely undeveloped, offering a quiet alternative to the busy Birmingham main line. Even the industrial stretches are often quiet and remote, enjoying the isolation imposed on them by the end of commercial traffic. Kestrels now hover above the waters of the Walsall Canal.

The guide does not cover all the BCN in the usual scale, and so details of many of the features that make up the BCN network are given alphabetically below.

Anglesey Branch

The branch leaves the Wyrley & Essington Canal at Ogley Junction, and runs north-west to Chasewater. It is predominantly a rural canal, passing through heathland before ending in the wide basin, overshadowed by the dam that contains the reservoir. Despite its rural nature, this canal saw the last regular coal traffic on the BCN, carrying coal from the Cannock mines until 1967. The course of the railway that used to serve the basin, and the remains of the loading wharves, can still be traced without difficulty. The branch is the ideal approach to the pleasure park that now surrounds the reservoir.

Bentley Canal

Built after 1840 as an additional link between the Walsall and the Wyrley & Essington Canals, little of the Bentley is now usable. Its eastern end is cut by a motorway, and survives mainly as a feeder for Walsall power station. The western end exists for six locks after Wednesfield Junction, but the rest of the canal has been converted into a water channel.

Cannock Extension Canal

Opened in 1858, the Cannock Extension was built to serve the coalmines of Cannock Chase. It was the last BCN canal to be built, and it was also one of the last to carry regular commercial traffic. As built it ran in a straight line to Churchbridge locks which linked it with the Staffs & Worcs Canal, but the northern half has vanished completely beneath the A5 and the neighbouring coalfields. It now serves Norton Canes Docks, a centre for narrow boat building and maintenance. Despite its recent industrial past, the Cannock Extension is totally rural; heath and woodland surround it, and there are few buildings to be seen. The water is clear, the towpath rich in wild flowers. The late building of the canal is revealed by its straightness, and by the distinctive blue brick bridges.

Dudley Canal

The Dudley Canal, whose history is intimately involved with that of the Stourbridge Canal, was built in several stages. Today it survives only in truncated form, but still serves as the vital link between the BCN and the west of England via the Staffs & Worcs Canal. The canal starts at the eight Black Delph locks, which were rebuilt in 1858. The old locks can be seen to the west of the present flight. The surroundings are largely industrial, but these are often exciting; the canal passes right through Round Oak steelworks at Brierley Hill. Parkhead locks, reopened in 1973, lead to Dudley Tunnel (see below). From Parkhead Junction the canal continues eastwards along a meandering contour line which takes it round Netherton Hill, dominated by the tower of Netherton Church. Much of the land around the canal is now heath and waste land, and so there are good views to the south. After the short Bumblehole branch, where one yard keeps alive a long tradition of boat-building, the canal divides at Windmill End Junction. The northern line leads to Netherton tunnel, the southern to Coombeswood Basin. Originally this line continued in a wide circle to join the Worcester & Birmingham Canal via the notorious Lappal tunnel, which was closed in 1917 because of subsidence. 3,795yds long, this rocky tunnel was the longest in the BCN network and one of the narrowest in the country. The short Gosty Hill tunnel (577yds) precedes Coombeswood Basin, the present end of the Dudley Canal.

Dudley Tunnel

Reopened to boats in 1973 with the rebuilding of Parkhead locks, Dudley tunnel is one of the wonders of the BCN. This narrow tunnel, 3,172yds long, was opened in 1792, after the usual delays and problems, to connect with the Birmingham Canal at Tipton. Inside the tunnel there is a vast network of natural caverns, basins and branches serving old quarries and mines. In all there are over 5,000yds of underground waterway, some cut off and abandoned, others still accessible.

Netherton Tunnel

Opened in 1858, Netherton was the last canal tunnel to be built in Britain, and the most luxurious. 3,027yds long, it was built with a bore sufficient to allow a towpath on both sides, and when opened it was equipped with gas lighting, later converted to electricity. The Netherton line joins the Birmingham main line at Dudley Port. The tunnel was built to relieve congestion in the Dudley tunnel, and runs on a parallel course.

Tame Valley Canal

Opened in 1844, the canal was built to overcome the long delays caused by the Farmer's Bridge locks. With towpaths on both banks, this straight canal leaves the Birmingham & Fazeley at Salford Junction, its north westerly course overshadowed at first by the motorway. As it climbs Perry Barr locks it comes into more open country, passing through the gradual stages of suburbia. There is a fine view of Birmingham from the top of the locks. At Rushall Junction the canal swings to the west, crossing the M5 motorway, the railway and the river Tame on one great embankment. Its elevated course continues as the surroundings become more industrial, and finally it joins the Walsall Canal at Tame Valley Junction, dwarfed by the cooling towers of Ocker Hill power station. The Rushall Canal was built in 1847 to connect the Tame Valley with the Daw End branch of the Wyrley & Essington, and thus to capture some of the coal trade from the Cannock mines. It continues the northern line of the Tame Valley Canal to Longwood Junction, the nine locks raising the canal through open country. The Rushall Canal is well known to anglers for its ample stocks of fish.

Titford Canal

Opened in 1837, the canal climbed originally to Causeway Green via six locks. Today it survives in shortened form, serving as a feeder from Rotton Park reservoir. The locks are still in good order, but much of the waterway is shallow.

Walsall Canal

The Walsall Canal runs from Ryder's Green Junction to Birchills Junction, making an alternative link between the Birmingham Canal main line and the Wyrley & Essington. It connects with the Tame Valley Canal and with the remains of the Bentley Canal. It was started in 1786 as a branch from the Birmingham line to serve Walsall, but did not reach Walsall until 1799. The link with the Wyrley & Essington was not made until 1841 because of company rivalry, but in that year the short Walsall Branch Canal was built to connect the two via eight locks. The course of the Walsall Canal is largely industrial, with a large number of basins, wharves and old arms. Despite the industry the canal is now quiet and remote, for much of its surroundings are now derelict and abandoned.

Wednesbury Old Canal

The specification for the original Birmingham canal included a branch to Wednesbury, to leave the main line at Pudding Green Junction. It was opened in 1769. Today it serves as a vital link between the Birmingham main line and the Walsall Canal, while its original course survives as the Ridgeacre branch, continuing eastwards towards West Bromwich.

Wyrley & Essington Canal

Opened throughout in 1797, the canal connected the Birmingham Canal with the Birmingham & Fazeley, running in a meandering contour line from Horseley Fields Junction to Huddlesford, via Lichfield. From Huddlesford it was able to connect with the Coventry, Oxford and Trent & Mersey Canals. The Wyrley & Essington was prompted by the coal trade, and there were several branches to serve the various coal fields. However its trade did not really develop until the Cannock fields were exploited in the 19thC. The most important was the Hay Head (or Daw End) branch, running southwardsfrom Catshill Junction, for this was later linkedwith the Tame Valley Canal via Rushall. The Wyrley & Essington was built as a rural canal, and today this is still largely the case. Most of its wandering course takes it through heath and woodland, although there are continual reminders of the industrial heritage of the area. The water is clear, and a great variety of plants grow on the towpath. In 1954 the main line between Ogley Junction and Huddlesford was abandoned, and much of this has now completely vanished, although its course through Lichfield and west of Huddlesford can still be traced. Today the canal ends abruptly at Ogley, but its connections with the Tame Valley and Walsall Canals, and with the Birmingham main line, make it an important part of the BCN network, a part that is emphasised by the rural nature of the canal.

Birmingham and the Black Country

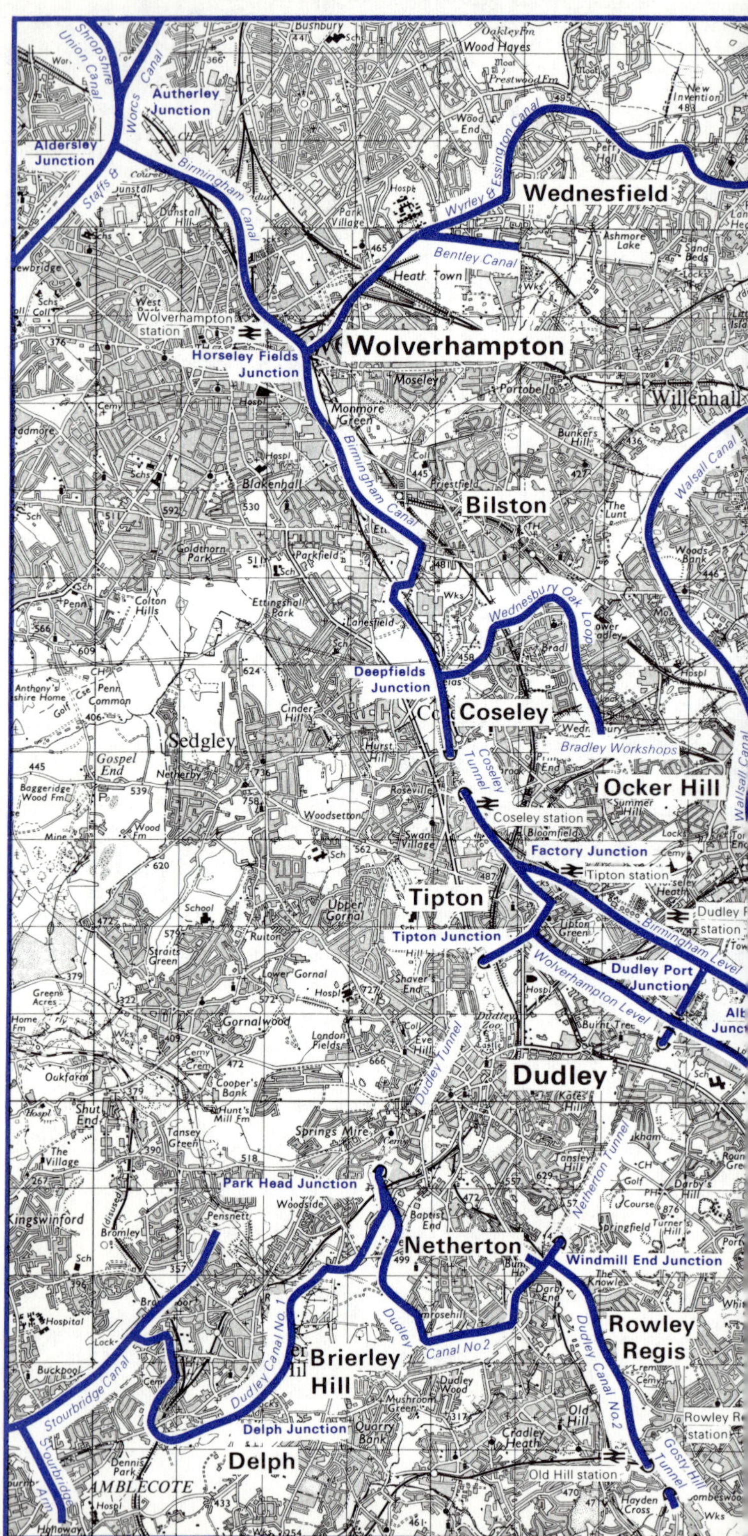

Daw End
Walsall
Wednesbury
West Bromwich
Perry Barr
Oldbury
Winson Green
Smethwick
Warley
Birmingham
Birchills Junction
Wyrley & Essington Canal
Walsall Branch
Walsall station
Daw End Branch
Rushall Canal
Rushall Junction
Tame Valley Canal
Bescot station
Tame Valley Junction
Ridgacre Branch
Ryder's Green Junction
Pudding Green Junction
Spon Lane Junction
Bromford Junction
Oldbury Junction
Titford Canal
Smethwick Rolfe Street station
Smethwick West station
Langley station
Smethwick Junction
Engine Branch
Soho Loop
Farmers Bridge Junction
Birmingham & Fazeley Canal
Icknield Port Loop
Rotton Park Reservoir
Oozells Street Loop
Worcester & Birmingham Canal
New Street station
Great Barr station
Perry Barr station

Birmingham and the Black Country

Geographically the Black Country is a large area of the Midlands stretching north west from Birmingham. Birmingham is not part of the Black Country, but could be called the root from which the whole area developed. The Black Country is a vast industrial conurbation, filling a square roughly bounded by Stourbridge, Wolverhampton, Brownhills and Birmingham. Many of the towns and villages swallowed by this square are old enough to have individual histories dating back to the medieval period and beyond. Although these are often very hard to trace today, the traditions die hard, and each town tries to preserve some vestige of individuality. The large scale industrial development is comparatively recent, for prior to the 18thC the area was still predominantly agricultural; since medieval times there had been some exploitation of mineral resources, iron, coal, limestone and clay, and the area was known for its lock, nail and buckle making, a localised and limited industrial development. The great change came in the 18thC when, following Abraham Darby's example at Coalbrookdale, the Midlands iron founders turned to smelting with coke instead of charcoal. All the necessary ingredients were available in the area in huge quantities, and so industrial growth was inevitably very rapid. Iron works and coal mines soon filled the landscape, and at the same time existing industries, for example the glass making at Stourbridge, underwent rapid expansion. The land and sky, darkened by smoke and industrial debris, gradually assumed the familiar character of the Black Country, and the foundations of 19thC prosperity were laid.

A vital part of the industrial development of the Black Country was the network of canals, well established by the end of the 18thC. These were built to ease communication within the area, and to connect it with markets throughout England and Europe. They provided the essential transport links which were the foundation of the prosperity of the Black Country.

Birmingham

Pop 1,107,000. EC Wed, MD Thur. Birmingham is not well known as one of the great tourist centres of Britain. Although the city has a medieval past—Smithfield market stands on the site of a moated manor, and the Bull Ring shopping centre is named after the habit of bull baiting—Birmingham is a memorial to the industrial revolution of the 18th and 19thC. Industrial and commercial development occurred so rapidly throughout the 19thC that the city soon became the trade centre of the Midlands, and of England generally. It still holds this position today, and extensive rebuilding programmes have made it one of the most modern cities in Britain, with an elaborate system of urban motorways that inevitably pay scant heed to anyone wanting to walk round the city. The growth of Birmingham during the late 18thC was aided by many famous industrialists and scientists who lived in the area, for example John Baskerville, William Murdock, Joseph Priestley, Matthew Boulton, James Watt.

Assay Office Newhall street. (021 236 6951). Collection of Birmingham and other silverware, coins, tokens and medals. Also Matthew Boulton's correspondence and books. *Open by appointment only.*

Aston Hall Frederick road. (2 miles N of city centre). (021 327 0062). A Jacobean house built in 1618–35, set in a fine park. Period furnishings in many rooms, and impressive friezes and ceilings. *Open Summer Mon-Sat 10.00–17.00, Sun 14.00–17.00 Winter Mon-Sat 10.00–dusk.*

Barber Institute of Fine Arts
The University, Edgbaston. (021 472 0962). Masterpieces by Bellini, Rembrandt, Rubens, Reynolds, Gainsborough, Courbet, Dégas, Monet, Whistler & others. *Open Mon-Fri 10.00–16.00, also 1st Sun afternoon of every month during university term.*

Birmingham City Museum & Art Gallery
Congreve street. One of the most important English provincial museums, with departments of art, ethnography, archaeology, natural history, science and industry. Fine Old Master paintings, and an exceptional collection of Pre-Raphaelite works. Pinto collection of wooden bygones, also industrial relics. *Open Mon-Sat 10.00–17.00, Sun 14.00–17.00.*

Museum of Science & Industry Newhall street. (021 236 1022). Exhibits of general scientific and industrial interest, including steam engines, motor vehicles, bicycles, machine tools, guns. *Open Mon-Sat 10.00–17.00, Sun 14.00–17.30.*

Sarehole Mill Colebank road, Hall Green. (021 777 6612). 18thC water powered corn mill, used also by Matthew Boulton for metal working. Restored to working order. *Open daily (including Sun) 14.00–19.00 (Sat 11.00–19.00). Closed winter.*

[i] TOURIST INFORMATION CENTRE

Birmingham Council House, Victoria square. (021 235 3411). Weekly information leaflet available, from here and from banks and libraries.

The development of the Black Country conurbation has inevitably merged many of the original towns and villages, and it is difficult to be aware of town boundaries while navigating the canals. Apart from Birmingham itself, there are many individual features of interest within the area. Some of these are listed alphabetically below. Where relevant, addresses and telephone numbers are given; but intending visitors should generally refer to the Tourist Information Centre for precise directions as access points to the canal network are very irregular.

Chasewater
The reservoir was originally a canal feeder, but it is now used for general water storage. The area has been developed for amenity and recreation, and there are facilities for sailing and boating. In 1967 the world's first 24 hour international speedboat race

on inland water was held here. Similar recreational facilities exist on Rotton Park reservoir, only half a mile from the centre of Birmingham.

Chasewater Park Railway Preservation Society

Brownhills (off A5). Large collection of relics, rolling stock and industrial locomotives. $2\frac{1}{2}$ miles of line. *Open from 14.00 Sat, Sun all year.*

Dudley

Dudley is a centre of heavy industry. One industry now extinct is nail making, but in the 19thC there were over 10,000 people in Dudley involved in this trade. The town sits astride a steep ridge, which is pierced by both the Netherton and Dudley tunnels. The Borough Council has given considerable support to the re-opening of Parkhead locks and the Dudley canal tunnel.

Dudley Castle and Zoo Only ruins of the 14thC castle survive. The fortifications were destroyed during the Civil War. The residential section survived until the 18thC, when it was largely destroyed by fire. In the 19thC the ruins were restored, and the grounds laid out as a public park. The zoo was opened in 1937, and contains a small mixed collection. *Open daily 10.00–18.00.*

Dudley Museum and Art Gallery St James' road. (56321). Permanent art collection and visiting exhibitions. The museum has an excellent geological section, and a Black Country Museum is under development. See below. *Open weekdays 10.00–18.00. Closed Bank hols.*

Black Country Museum This exciting project, backed by Dudley Borough Council, is being developed on a site near Dudley town centre with access from the A461; it is adjacent to the present museum and art gallery, and overlooks the entrance to Dudley canal tunnel. The aim is to capture and preserve exhibits which illustrate the industrial, economic and social development of the Black Country. These range from machine tools and colliery equipment to canal boats and trams. There will also be facilities for lectures and films. Further details from: The Director, Dudley Museum and Art Gallery, St James' road, Dudley, Staffs. (56321).

Wren's Nest Geological Nature Reserve Dudley, over the canal tunnel. Two trails have been laid out over 74 acres of upper silurian rock, an area rich in fossils. Information from the Dudley Museum and Art Gallery.

Soho Foundry, Smethwick

In 1796 Matthew Boulton and James Watt built the famous Soho foundry beside the canal in Smethwick, opposite the Soho manufactory which Boulton had established several years before. They were joined by William Murdock, who developed a form of gas lighting which was installed in the foundry early in the 19thC. Here was carried out the production of steam engines, boilers, gas apparatus and other heavy industrial goods. Boulton was already well known as a silversmith and maker of steel and silver plate, and so his connection with Watt played a vital role in the industrial development of the area. Later in the 19thC a mint was installed in the Soho foundry. Many examples of the work produced at Soho can be seen in the Birmingham Museum of Science and Industry. A Boulton and Watt engine can be seen in operation periodically at Crofton, on the Kennet and Avon canal (see page 33).

Avery Historical Museum Soho Foundry (021 558 1112). Machines, instruments, weights and records illustrating the history of weighing. *Open by appointment only.*

Walsall

Apart from heavy industry, Walsall has long been renowned as a centre of the leather trade which developed in close relationship to the manufacture of saddlers' ironmongery. Nearby Willenhall was famous for the manufacture of locks and keys. In 1855 there were 340 locksmiths working in and around the town.

Museum of Leathercraft Central Library and Art Gallery, Lichfield street. (23920). Permanent exhibition illustrating the past, present and future of the leather trade. In the same building is the E. M. Flint Art Gallery, housing the Jerome K. Jerome collection. *Open daily 10.00–19.00, Sat 10.00–17.30.*

Rough Wood Nature Trail Mining subsidence in the area has prompted the growth of plants more characteristic of bogs and marshes. The trail has been laid out to illustrate these. Information from Parks Dept., Council House, Walsall.

West Bromwich

The town has recently undergone drastic rebuilding and redevelopment, and has expanded to swallow its suburbs, Tipton and Wednesbury, the latter named after Woden, the norse god of pre-Christian times. St Bartholomew's church, Wednesbury, dates back to the 12thC, but was heavily restored in the 19thC; it contains fine Victorian stained glass.

Oak House Museum Oak road. (021 553 0759). 15thC timber-framed house furnished in period style. *Open weekdays 11.00–17.00, Sun 14.30–17.00.*

Wolverhampton

This dull town has grown up around the heavy industries which are its heart and its wealth. Extensive redevelopment recently has produced the Mander centre and the Wulfrun shopping precinct, the latter named after one of the founders of Wolverhampton in the 7thC. Little survives of any age in the town, but the church of St Peter dates back to the 15thC.

Municipal Art Gallery and Museum Lichfield street. (24549). Collections of English enamels, Staffordshire pottery, and English painting and sculpture. *Open weekdays 10.00–18.00.*

Bantock House Bantock park. (24548). Collection of enamels, English porcelain and Chinese ivories, dolls and costume. *Open weekdays 10.00–19.00, Sun 14.00–17.00.*

Bilston Museum and Art Gallery Mount Pleasant, Bilston. (42097). The museum illustrates the life and activities of Bilston and the Black Country, with an emphasis on the enamels made here in the late 18thC and early 19thC. (Replicas of these enamels are now made in Battersea, London.) *Open weekdays 10.00–17.00.*

RESTAURANTS

XY **Burlington Restaurant** Burlington Passage, New street, Birmingham. (021 643 3081). Basement restaurant offering a mixture of English and continental food. Good Food Guide since 1967.

XY **Copper Kettle** 151 Milcote road, Bearwood, Birmingham. (021 429 7920). Excellent French-inspired menu. Good Food Guide 1973. *Closed Sun, and all Aug.*

XY **La Capanna** Hurst street, Birmingham. (021 622 2287). Italian food, imaginative and ever changing menu. Good Food Guide since 1969. *Closed Sun and B. Hols.*

XY **Danish Food Centre** Stephenson Place, Birmingham. (021 643 2837). Useful Danish restaurant, open all day, from 8.00–22.30. Good mixture of hot and cold dishes, draught lager. *Closed Sun, Mon dinner.*

X **Long Boat** Cambrian wharf, Kingston row. Canalside. Steak bar.

X **Loon Wah** 11 Bromsgrove street, Birmingham. (021 622 2697). Basic but sound Cantonese cooking. Good Food Guide since 1969.

XY **Salamis Kebab House** 178 Broad street, Birmingham. (021 643 2997). Greek-Cypriot restaurant, near city centre. Customers choose their food in the kitchen.

XY **Rendezvous Restaurant** 40 Berry street, Wolverhampton. (23481). Olde worlde restaurant, with furniture and staff to match. Mixed English and continental menu.

BOATYARDS & BWB

Ⓑ **C. J. Parker** Norton Canes Dock, Lime Lane, Pelsall, Staffs. (Brownhills 3179). On the Cannock Extension Canal. Water, dry dock and slipway, boatbuilding, sales and repairs.

Ⓑ **Canal Transport Services** Norton Canes Docks, Lime lane, Pelsall, Staffs. (Brownhills 4370). On the Cannock Extension Canal. Water, diesel, gas, chandlery, refuse and sewage disposal, drydock and slipway, moorings, winter storage, boatbuilding, sales and repairs, inboard and outboard engine sales and repairs. 50-seater trip boat for hire to parties, motor boats for hire by the hour.

Ⓑ **M. E. Braine** Norton Canes Docks, Lime lane, Pelsall, Staffs. (Brownhills 4888). On the Cannock Extension Canal. Water, diesel, petrol, gas, chandlery, refuse disposal, dry dock and slipway, moorings and winter storage, boatbuilding and conversions, inboard engine sales and repairs. 46-seater trip boat for hire to parties.

Ⓑ **Bumblehole Boating Centre** St Peter's road, Netherton, Dudley, Staffs. (Dudley 55874). Bumblehole branch by Netherton tunnel. Hire cruisers, water, diesel, petrol, gas, chandlery, dustbins and sewage disposal, slipway, moorings, winter storage, boat building, sales and repairs, inboard and outboard engine sales and repairs.

Ⓑ **Alfred Matty** Biddings lane, Deepfields, Coseley, Staffs. (Bilston 42725). On the BCN main line. Water, crane available, emergency repairs.

Ⓑ **James Walton & Son** Biddings lane, Deepfields, Coseley, Staffs. (Bilston 42725). Traditional and modern canal boats built to order, in elm or oak.

BWB Canal Shop & Information Centre 2 Kingston row, Birmingham. (021 236 4844). Chandlery, canal publications and wares for sale, general BWB information and leaflets available. Moorings, water, refuse and sewage disposal at Cambrian Wharf administered from here. *Closed Sun.*

BWB Permanent Moorings available at Longwood, at the head of the Rushall locks. (And at Cambrian Wharf, Birmingham.)

FISHING

Inevitably much of the water of the BCN is dirty, and in some places fish find life rather a struggle. However, in the more rural parts of the network the water is clearer, and not nearly so infested with rubbish and débris. In this more hospitable environment fish thrive.

The Tame Valley Canal, the Rushall Canal and the Wyrley & Essington Canal are all available to the general public for fishing, and can be used on day tickets, or for match fishing. All the usual varieties can be caught. In January 1973 the Rushall Canal was restocked with 6,000 fish, as part of the BWB fishery management policy.

Season permits, day tickets etc. available from the BWB water bailiff: W. Bust, Lock House, Parkhead Locks, Park Hall Road, Walsall, Staffs.

The Long Boat pub at Cambrian Wharf, near the centre of Birmingham. The narrowboat on the right houses a bar.

Stratford -on-Avon

Maximum dimensions

Length: 70′
Beam: 7′
Headroom: 6′

Mileage

KING'S NORTON JUNCTION to
Hockley Heath: 9¾
LAPWORTH junction with Grand Union Canal: 12½
Preston Bagot: 16¼
Wootton Wawen Basin: 18½
Wilmcote: 22
STRATFORD-ON-AVON junction with river Avon: 25½

Total 55 locks

The opening of the Oxford Canal in 1790 and of the Coventry Canal throughout shortly afterwards opened up a continuous waterway from London to the rapidly developing industrial area based on Birmingham. It also gave access, via the Trent & Mersey Canal, to the expanding pottery industry based on Stoke-on-Trent, to the Mersey, and to the East Midland coalfield. When the Warwick & Birmingham and Warwick & Napton canals were projected to pass within 8 miles of Stratford-on-Avon, the business interests of that town realised that the prosperity being generated by these new trade arteries would pass them by unless Stratford acquired direct access to the network. And so, after the usual preliminaries, on the 28th of March 1793 an Act of Parliament was passed for the construction of the Stratford-on-Avon Canal, to start at King's Norton on the Worcester & Birmingham Canal (itself a long way from completion at that time). The junction was to be less than 3 miles from the junction of the Worcester & Birmingham with the Dudley Canal, and would thus provide a direct route to a major coal producing area without passing through Birmingham.

Progress was rapid at first; but almost the total estimated cost of the complete canal was spent in the first three years, on cutting the 9¾ lock-free miles to Hockley Heath. It took another four years, more negotiations, a revision of the route and another Act of Parliament to get things going again. By 1803 the canal was open from King's Norton Junction to its junction with the Warwick & Birmingham Canal (now part of the Grand Union main line) near Lapworth, with through traffic along the whole of this northern section. Even more delays now followed, with little enthusiasm on the part of private investors to put up more money. Cutting recommenced in 1812, the route being revised yet again in 1815 to include the present junction with the river Avon at Stratford. (From here, the Avon was navigable down to the Severn at Tewkesbury.)

In its most prosperous period, the canal's annual traffic exceeded 180,000 tons, including 50,000 tons of coal through the complete canal, down to Stratford. By 1835 the canal was suffering from railway competition. This grew so rapidly that in 1845 the Canal Company decided to sell out to the Great Western Railway. There was opposition, however, and it was not until 1st January 1856 that the sale was considered complete. Thus the canal had been in full, independent operation for less than 40 years. Traffic was not immediately suppressed by the new owners, but long distance haulage was the first to suffer as it was a more direct threat to the railway. In 1890 the tonnage carried was still one quarter of what it had been 50 years before, but the fall in ton-miles was much greater.

This pattern of decline continued in the 20thC, and by the 1950s only an occasional working boat was using the northern section; the southern section (Lapworth to Stratford) was badly silted, some locks were unusable and some of the short pounds below Wilmcote were dry. It is believed that the last boat to reach Stratford did so in the early 1930s but there is evidence that a pleasure cruiser reached Wilmcote during the Easter holiday of 1947.

After the 1939–45 War interest began to grow in boating as a recreation. In 1955 a Board of Survey had recommended sweeping canal closures, including the southern section of the Stratford canal. Public protest was such that a Committee of Enquiry was set up in 1958, and this prompted the start of a massive campaign to save the canal. The campaign was successful: the decision not to abandon the canal was announced by the Ministry on the 22nd of May 1959. On October 16th of the same year the National Trust announced that it had agreed a lease from the British Transport Commission under which the Trust would assume responsibility for restoring and maintaining the southern section. The transfer took place on the 29th September 1960 and restoration work began in earnest in March of the following year. The terms of the arrangement included a contribution towards the cost of restoration but a very substantial sum was provided by the Trust, which now maintains the southern section at its own expense. (The freehold was transferred by BWB to the Trust in 1965.)

The re-opening ceremony was performed by Queen Elizabeth the Queen Mother on the 11th of July 1964, after more than four years of hard work by volunteer prisoners, canal enthusiasts, Army units and a handful of National Trust staff. This magnificent operation has acted as an encouragement to all other would-be canal restorers since then. It represents a triumph for volunteer effort.

Natural history

Between the green tunnel of trees near King's Norton and the lily-covered basin at Stratford, there is great scope for seeing the wild life and plants that frequent this delightful Warwickshire canal. After passing through Shirley, the canal enters the old Forest of Arden and for miles the banks are bordered by sturdy oaks and hazel bushes which are a great attraction to grey squirrels. Here, the harsh calls of jays and the rattle of magpies are familiar sounds. It is worth making a halt at Earlswood, for it is only a few minutes' walk to the feeder reservoirs which attract naturalists from all over the Midlands. Here, at any time of the year, several pairs of great crested grebes can be seen carrying out their curious courtship ritual; the birds face each other, shake their heads from side to side, then dive and present offerings of weed. In early summer the young grebes can sometimes be seen riding on the parents' backs.

East of the open fields of Hockley Heath, the canal descends through the Lapworth flight. The locks and bridges are interesting as the walls are often covered with a profusion of plant life, including the small ferns of the spleenwort family. Further down, in the southern section past lock 29, there is a magnificent display of hartstongue ferns growing near a demolished railway bridge. Past lock 27, the western bank is lined with large alders and pollarded willows, the haunt of tits and warblers. During late summer these trees provide a roosting place for countless swallows.

At Yarningale, the canal crosses a tributary of the River Alne, which flows parallel for several miles. The dipper, which is rather like a blackbird but with a white breast, is sometimes seen around here, skimming over the water or perched on a stone. It is unique in being able to 'fly' underwater, as well as walk on the stream bed in search of insects. From Preston Bagot to Bishopton, the canal passes through one of the richest parts of Warwickshire as far as wildlife is concerned. Badgers are very common in the area and can sometimes be observed from the canal several hours before sunset; and with luck, a weasel can be seen hunting on the banks. These animals can be observed only when the boat is moored and quiet. It is worth a pause at bridge 51 and a walk up the bridle path to Austy Wood. Apart from numerous warblers, this wood houses a large number of great spotted woodpeckers whose drumming can be heard during spring. In this area, Canada geese and several species of ducks frequently fly over to Wootton Pool. The harsh 'kwark' of the heron is a familiar sound, for there is a small heronry at the Pool and the birds are busy at their nests from March until August. At least four pairs of kingfisher have their homes on the Stratford canal, and they may occasionally be seen perched on lock gates as they watch for rising fish. Another interesting bird, which can be heard singing during late afternoon and evening, is the grasshopper warbler; it sounds just like a grasshopper or a fisherman's reel. The moorhen is the most common water-bird on the canal; 57 have been counted in the five mile stretch between Lowsonford and Wilmcote. The moorhen nests from early spring to late September, so great care should be taken to avoid clumps of reeds in which a nest may be concealed. Many nests are easy to spot from a boat and, as a rule, they are limited to one side of the canal—opposite the towpath. Moorhens feed on almost anything; they can be seen pecking at blackberries, pulling leaves of plants or foraging in the fields amongst cows. At night they frequently roost in the alder trees lining the bank.

Wild flowers of the canal are too numerous to list. Purple loosestrife is one of the most showy flowers but is not found above Wootton Wawen. It is one of the food plants of the elephant hawkmoth caterpillar, a creature some three inches long with a pair of conspicuous eye spots near the front of its body. The arrowhead is generally considered to be quite rare in the British Isles, but towards Stratford its peculiar arrow-shaped leaves and spikes of white flowers are abundant in summer. The flowering rush is fairly common throughout the canal. Anyone particularly interested in pond life should dip his net at the winding holes where all manner of aquatic insects are to be found among the pink spikes of bistort.

Enquiries concerning the natural history of the Stratford-on-Avon Canal may be addressed to the Department of Natural History, the City Museum and Art Gallery, Birmingham, B3 3DH.

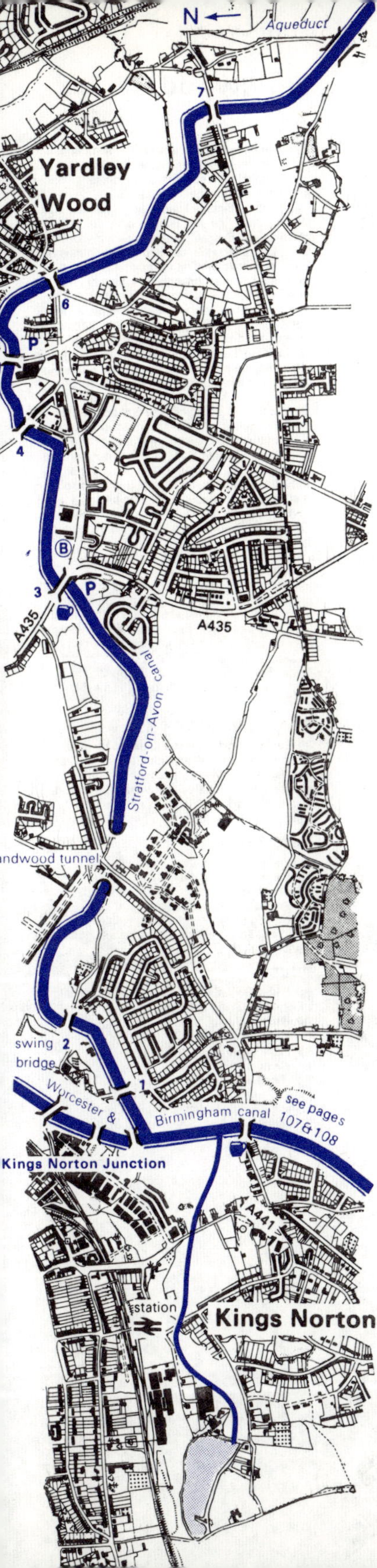

King's Norton

$4\frac{1}{4}$ miles

The west end of this delightful canal is at King's Norton, just outside Birmingham. As may be guessed from the map, the first 5 miles of the navigation pass entirely through residential outskirts of the Birmingham conurbation. The canal veteran might expect this to be a dull or scruffy stretch, but he would be wrong: in fact the Stratford canal is bordered all the way with dense but varied vegetation and, thus protected from the inroads of the suburbs, forms a quiet, winding ribbon of green all the way through to the real countryside. In conjunction with the northern section of the Worcester & Birmingham Canal, this is a far more scenically interesting route between Lapworth and Birmingham than via the Grand Union Canal.

Leaving the Worcester & Birmingham (*see page 107*) at King's Norton Junction, the Stratford-on-Avon canal proceeds straight to the well-known King's Norton stop lock. In the days of the private canal companies, stop locks were common at junctions, as one canal sought to protect its water supply from any newcomer; but King's Norton stop lock is unusual in having 2 wooden guillotine gates mounted in iron frames balanced by chains and counterweights. The machinery is now seized up, and boats pass under the 2 gates without stopping. The next bridge is a small swing bridge, then round the corner is Brandwood tunnel. Further east is a beautiful tree-lined cutting, then a bridge with a pub beside it (*petrol & telephone nearby*) and the remains of an old arm just beyond it. Round the corner is a private boatyard; a water point is outside a cottage near bridge 5. The canal continues through pleasant wooded cuttings – access is bad at most of the bridges, so the seclusion is virtually complete.

Brandwood Tunnel
352yds long, this tunnel has no towpath. Horse-drawn boats had to be hauled through by means of an iron hand-rail on the side. Lengths of this rail can still be seen in the tunnel.

PUBS

Horse Shoe Canalside, at bridge 3. Cold food. *Telephone and petrol nearby.*

FISHING THE STRATFORD-ON-AVON CANAL

The Stratford-on-Avon canal from King's Norton to Stratford is a popular fishing water, holding many species of coarse fish. The canal has a number of pounds throughout its length, and these pools make favourite swims of the local anglers. At Wootton Waiven there are good tench, bream and some interesting fishing for the crucian carp – (the latter also at Stratford). Lapworth is a good place for roach, with several fish over the 1lb mark to be caught most seasons. Throughout the length there are roach, rudd, perch, tench and gudgeon. There are two varieties of carp to be encountered – the smaller crucian carp, and some common carp of around 5lb and 6lb.

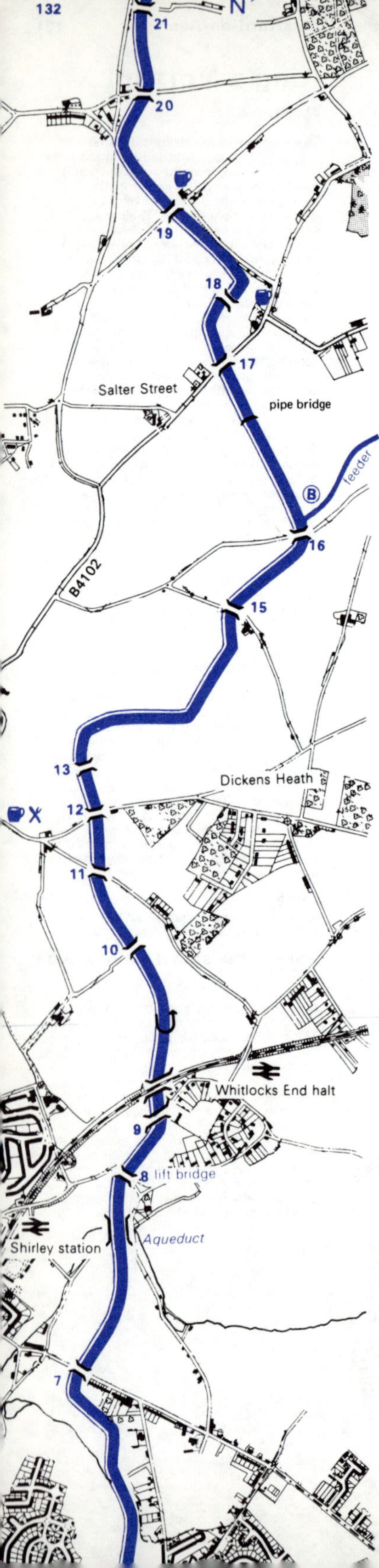

Earlswood

$5\frac{1}{4}$ miles

Passing over a small aqueduct, the canal reaches a steel lift bridge, which has to be raised and lowered with a windlass. (There are 2 more of these bridges nearer Lapworth). Beside the bridge a former pub, the Boatman's Rest, is now only an off-licence. Passing under a railway bridge, the canal sheds all traces of the suburbs but maintains its twisting course in wooded cuttings through quiet countryside. The bridges over the navigation are mostly the generous brick arched bridges typical of the canal between King's Norton and Lapworth locks (in contrast to the much smaller bridges further south), but few roads of any significance come near the canal. At. bridge 16 the canal emerges from a long cutting and is joined by a feeder from the nearby Earlswood Reservoirs. Several boats are moored along it, for at the junction there is a boatyard and the mooring site of the Earlswood Motor Yacht Club. There are no villages along this rural stretch of canal, but at Salter Street there is a modern school and a strange Victorian church. At bridge 19 (which spans a cutting) is a cider house. One of the buildings here was once a brewhouse.

Earlswood Reservoir
Half a mile south of bridge 16 is this canal-feeding reservoir, surrounded by trees and divided into three lakes: Windmill Pool, Engine Pool and Terry's Pool. There is a sailing club on Terry's Pool, and fishing available to the general public on the other two. The BWB bailiff will supply information about the lakes. (Lapworth 2091).

BOATYARDS & BWB

Ⓑ **Earlswood Marine Services** Lady lane, Earlswood, Warwicks. (2552). Diesel, gas, chandlery, sewage disposal. Slipway, moorings. Boat sales and repairs, inboard and outboard engine sales and repairs. Large trip boats for hire. *Open all year.*

BOAT TRIPS

'Franklin', 'Planet' and 'Cepheus' are full length motor narrowboats available for party bookings from Earlswood. Each has a bar and a piano on board, and carries a maximum of 45 to 50 passengers. All enquiries to Earlswood Marine Services (*see above*). *April-October only.*

PUBS

Blue Bell Cider house. Canalside, at bridge 19. Good mooring jetty.
Bull's Head $\frac{1}{4}$ mile S of bridge 17. Old country pub.
Red Lion 500yds S of bridge 16, near Earlswood Reservoir.
Three Maypoles restaurant. Tamworth lane, Shirley (1905). $\frac{1}{4}$ mile NE of bridge 12. Lunch *12.00–14.00 Mon-Fri,* Dinner *19.00–22.00 Tue-Sat. Closed Sun.*

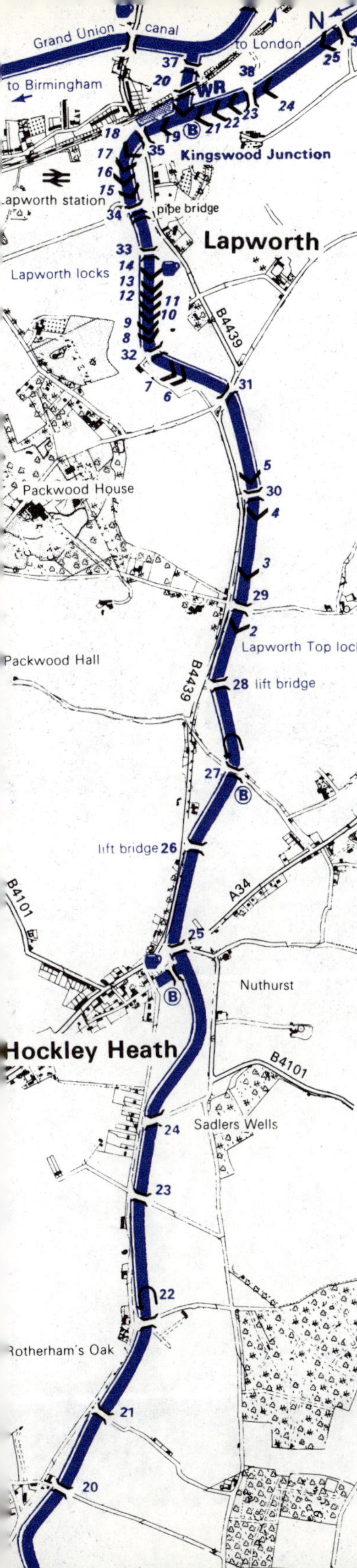

Lapworth locks

4¾ miles

The canal continues to wind gently south-eastwards through the quiet countryside, sometimes in minor cuttings and often flanked by trees. To begin with, there are no locks, and the bridges—especially those in the cuttings—are still the big brick arches worthy of a broader canal. At Hockley Heath (bridge 25) there are plenty of boats, for a tiny arm that once served a coal wharf is now a boatyard (Swallow Cruisers). The old stable is now used as a store. Next to the boatyard, the Wharf Inn overlooks the canal. Further along is another boatyard, catering for small fibreglass cruisers, flanked by 2 windlass-operated steel lift bridges. East of here things change dramatically, for the first of the 55 narrow locks down to Stratford is reached. (The top lock is numbered 2, the old stop lock at King's Norton being Number 1). The surroundings of the top lock are indeed pleasant: a white house surrounded by walls and hemmed in by trees stands beside the lock, while a cottage with a delightful garden faces the towpath just below. To the south-west can be seen the spire of Lapworth church. After the first 4 locks is a half-mile breathing space, then the Lapworth flight begins in earnest, with each of the next 9 locks spaced only a few yards from its neighbour. The short intervening pounds have been enlarged to provide a bigger working reservoir of water, so that one side of each lock is virtually an isthmus. The locks have double bottom gates and are not heavy going. They are interspersed with the old cast-iron split bridges that are such a charming feature of the Stratford-on-Avon canal. These bridges are built in two halves, separated by a 1-inch gap so that the towing line between a horse and a boat could be dropped through the gap without having to disconnect the horse. Below lock 19 is Kingswood Junction: boats heading for Stratford should keep right here. A short branch to the left through lock 20 leads under the railway line to the Grand Union Canal. (*See book 1 in this series*).

Lapworth
Warwicks. PO, tel, stores, garage, station. Indivisible from Kingswood, this is more a residential area than a village. Two canals pass through Lapworth: the heavily locked Stratford-on-Avon canal and, to the east, the main line of the Grand Union canal. These two waterways, and the short spur that connects them, are easily the most interesting aspect of Lapworth. There are usually plenty of boats about, especially on the southern section of the Stratford canal (below lock 21), the canalside buildings are attractive and there are 2 small reservoirs at the junction. The mostly 15thC church is quite separate from the village and is 1½ miles west of the junction; it contains an interesting monument by Eric Gill, 1928.

Packwood House
NT Property. Lapworth (2024). ½ mile N of the B4439 road bridge. A 16thC timber-framed house, enlarged in the 17thC by John Fetherstone. It was he who created the clipped yew garden that is held to represent the Sermon on the Mount. The house contains collections of tapestry, needlework and furniture. *Open Apr-Sep Tue, Wed, Thur, Sat, Sun and B Hols 14.00–19.00; Oct-Mar Mon, Wed, Sat, Sun, B Hols 14.00–17.00.*

Hockley Heath
Warwicks. PO, tel, stores, garage. A featureless place, but the several shops are conveniently close to the canal bridge.

BOATYARDS & BWB

Ⓑ **Narrowboat Services** Lapworth Basin, Lapworth, Warwicks. (2727). Wooden boat building; all types of boats fitted out and repaired. Crane. Drydock for full-length boats (ready early 1973).

Ⓑ **E. G. Wheatley** Old Warwick road, Lapworth, Warwicks. (Lapworth 2379).

Beside lock 14. Diesel, gas, moorings, winter storage. Groceries, hire cruisers.
Ⓑ **Holwill's Boatyard** Wharf lane, Hockley Heath, Warwicks. (Lapworth 3442). Beside bridge 27. Water, gas, chandlery, dustbins, sewage disposal. Slipway, moorings. Dealers in fibre-glass boats, inboard and outboard engines sales & repairs. *Open daily.*
Ⓑ **Swallow Cruisers** Rear of Wharf Inn, Stratford road, Hockley Heath, Warwicks. (Lapworth 2418). New and used boat sales, outboard engine sales and repairs. Gas, chandlery, slipway.

PUBS

Navigation Lapworth. Canalside, on Grand Union canal at bridge 65.
Boot Inn Lapworth, near lock 14.
Wharf Inn Hockley Heath. Canalside, at bridge 25. Food; and a garden down to the canal.

Good traditional decoration at Lapworth.

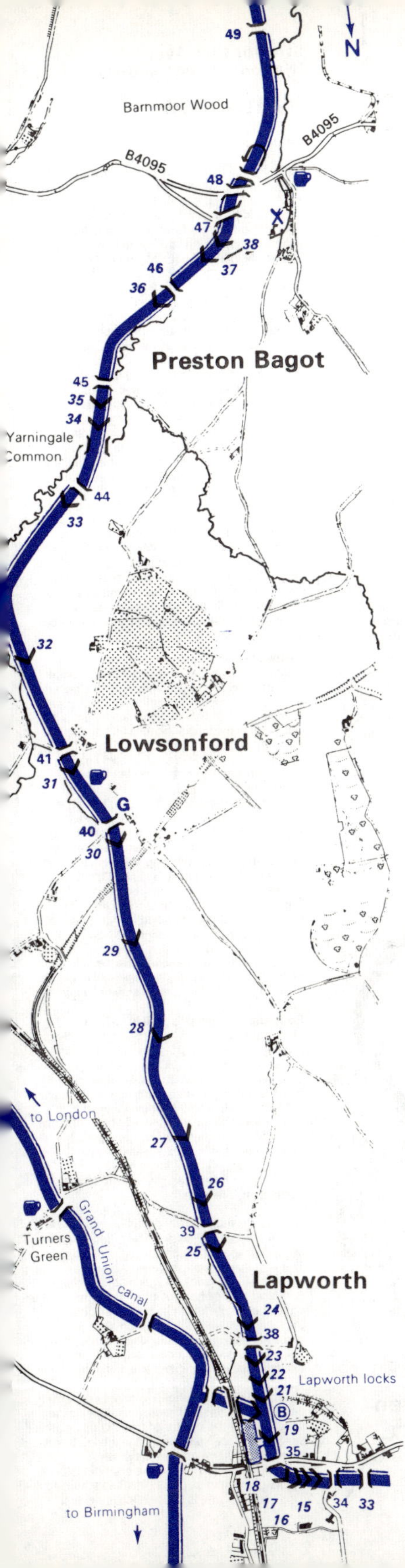

Preston Bagot

4½ miles

Navigational Note
The southern section of the Stratford-on-Avon Canal—i.e. from Lapworth to Stratford—belongs to the National Trust, not the British Waterways Board. The Trust has owned the canal since it was restored from dereliction and reopened in 1964 by Queen Elizabeth the Queen Mother. A special licence to navigate the canal must be obtained from the Trust's representative in the Canal Office by lock 21; a map and a very detailed guide book may also be bought here. Facilities for boats are also available.
At Kingswood Junction the Stratford Canal continues south, locking steadily downward. The locks have single gates and very small paddles, so they are slow to fill and empty. At some of them are the little iron split bridges over the lock-tail; and at intervals may be seen the delightful barrel-roofed lock-cottages which are just as much a hallmark of this canal as are the bridges. Most of these cottages are inhabited and well cared for—a great contrast to many of the cottages on the canal system. Indeed the whole of the southern Stratford has the appearance of a canal that is cherished by the people living along it—the towpath is in places beautifully mown, the hedges trimmed and the locks and bridges painted. The canal goes through folds of very pretty, wooded country, rarely encountering a main road or more than a scattering of houses. This virtual seclusion is maintained right through to Stratford. At Lowsonford is a pub where boats may moor in among the weeping willows of the garden; at Yarningale the canal, having followed a small stream for several miles, crosses it on a tiny iron aqueduct adjoining lock 34. Bread and ice cream can sometimes be bought at a cottage 100yds away; but down at Preston Bagot delicious home-made cakes are available.

Preston Bagot
Warwicks. A small, scattered settlement with attractive, ancient houses here and there, including the 16thC manor. The church of All Saints has a Norman nave and other Norman details, with Victorian additions.

Lowsonford
Warwicks. PO, tel, stores (all just west of lock 30). Another small and scattered hamlet, tucked away by the canal.

BOATYARDS

National Trust Canal Depot Lapworth Junction (3370). Water, refuse and sewage disposal, diesel, canal maps and guides. This is the Canal Manager's office and the main telephone number for all enquiries about the southern section of the Stratford-on-Avon canal.

PUBS

Old Crab Mill Preston Bagot, 350yds W of the new road bridge. (Claverdon 2857). *Telephone outside.*

Mill Cottage Tea Room Preston Bagot, 300yds W of the new bridge. Light lunches, teas, etc. Home-made cakes. *11.00–20.00 daily except Mon.* (Claverdon 2778).

Fleur de Lys Lowsonford, by the canal north of lock 31. Ancient building with large garden. (Lapworth 2431). Food.

Ye Olde New Inn Turner's Green. Canalside, on the Grand Union canal. (½ mile SE of Stratford Canal bridge 39, beyond the railway).

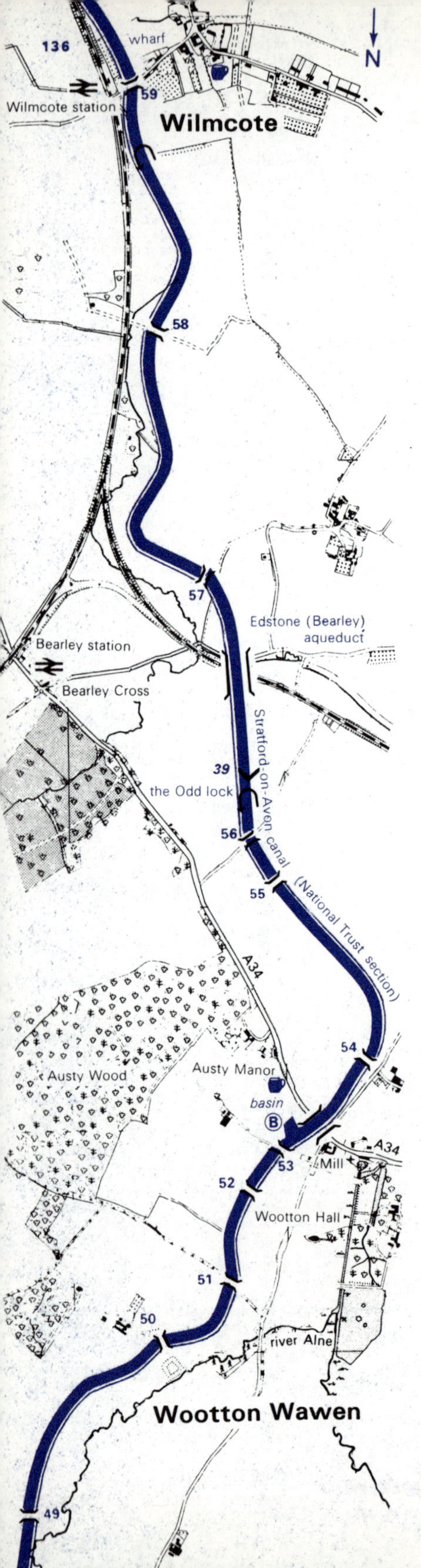

Stratford-on-Avon (National Trust section)

Wilmcote

5 miles

Lock 38 at Preston Bagot introduces two long pounds, which are welcome on such a heavily locked canal. The canal continues through delightfully quiet country, passing a farm as the big Austy Wood looms up on the hill to the east. (The low stone hall is Austy Manor). Beyond bridge 53 the canal widens into a basin—a boatyard and a pub are here—and then crosses the A34 road on a cast iron aqueduct. Soon it enters a slight cutting—rare on the southern section of this canal—and then straightens out at lock 39—known as the Odd lock. Further south the canal rises on an embankment and is then carried across watermeadows, a road and a railway by the splendid Edstone aqueduct. At the south end is a very pretty cottage and garden. The navigation now winds along a secluded course to Wilmcote. Just north of the village are the remains of a bridge—this used to carry a horse tramway that served nearby quarries. The winding hole and the cottages on the towpath at this point were built for the quarry trade. Wilmcote is close to the canal. Long- or short-term moorings at Wilmcote can be had at the private wharf just south of bridge 59, but boatmen must first ask the owner, Mr P. J. Plater, for permission. Mr Plater also sells bottled gas and has a slipway into the canal (Stratford-on-Avon 3998).

Wilmcote
Warwicks. PO, tel, stores, garage. A small and attractive village, typical of this part of the world. A beautiful lime tree on the green is the centre of the village: nearby are a fine old pub, a residential hotel and the most well-known building in the village—Mary Arden's cottage. The school and vicarage by the church were built by William Butterfield in *c.* 1845. The little railway station is to the east of the canal: with its trim roses and well-painted structures it is kept very much in the old tradition by its staff of former Great Western Railway men.
Mary Arden's Cottage Wilmcote. This was the home of Shakespeare's mother, and is a beautiful 15thC timbered farmhouse. The long, low house crouches behind luxurious and well-tended flower beds, and contains a museum of agricultural implements and local rural bygones. It is owned by the Shakespeare Birthplace Trust: *open Apr-Oct weekdays 9.00–18.00, Sun 14.00–18.00. Nov-Mar weekdays only, 9.00–16.00.*
Edstone (or Bearley) Aqueduct
This major aqueduct, approaching 200yds in length, consists of a narrow cast iron trough carried on brick piers across a shallow valley. As with the 2 other—but much smaller—iron aqueducts on this canal (at Yarningale and Wootton Wawen) the towpath runs along the level of the bottom of the tank, so that towing horses and pedestrians get a duck's eye view of passing boats. This feature makes the aqueducts on this canal very unusual.
Wootton Wawen
Warwicks. PO, tel, stores, garage, station. This scattered but very pretty village is half a mile west of the basin. The village has been designated a Conservation Area. There are plenty of timbered houses and the late 17thC Hall looks superb across the parkland and pond; but the chief glory is the church of St Peter on its rise overlooking the whole village. This church should certainly be visited. Its unusual building history has given it a pleasantly disorderly external appearance, but inside there are really rare and fascinating things to see. The church is the only one in Warwickshire that derives from Saxon times, and the original sanctuary in the centre of the 11thC church survives intact, still the focus of the church after over 900 years. The nave is conspicuously Norman, the chancel is bare but large, with a superb 14thC east window. The Lady Chapel is probably the oddest

part of the whole building – it is like a barn in more ways than one. It is enormous, with a primitive tiled roof and a completely irregular brick floor. Birds are often to be found, enjoying the shelter it provides. All around the walls is a medley of monuments. It all adds up to an intriguing building.

Wootton Wawen Basin
This wide, embanked basin was built when construction of the canal was halted here for a while. A hire cruiser base – one of the few on this canal – has been built here, and was in 1972 awarded a Civic Trust commendation for its design. With a nearby pub and petrol station, the wharf is a popular halt with both boatsmen and motorists. A cast-iron aqueduct carries the canal over the A34 by the basin. This aqueduct has been damaged twice in recent years by lorries hitting the underside, so now a triangular road sign warning motorists of the headroom is mounted on the aqueduct. Unfortunately this sign has been so positioned as to obscure almost completely the original iron plaque that commemorates the opening of the aqueduct in 1813. Just down the hill from the aqueduct is a fine brick watermill, in good repair. This dates from the late 18thC.

BOATYARDS

Ⓑ **Anglo Welsh Narrow Boats** The Wharf, Wootton Wawen, Solihull, Warwicks. (Henley-in-Arden 3427). Hire cruisers. Water, diesel, sewage disposal.

PUBS

Mason's Arms Wilmcote. Snacks and hot meals *daily except Sun lunch and Mon dinner.* (Curry a speciality). Permanent exhibition of paintings by the landlord's wife. (Stratford-on-Avon 2166).

Swan House Hotel Wilmcote (Stratford-on-Avon 2030). This converted Georgian house is close to Mary Arden's cottage, 400yds from the canal. Enterprising French food is served in a galleried dining room. The short menu changes every 3 weeks. *Lunch 12.30–13.30, dinner 19.30–20.45:* booking advisable. In the Good Food Guide since 1971.

Olde Bull's Head Wootton Wawen, ½ mile W of the aqueduct. Full restaurant meals daily. (Henley-in-Arden 2511).

Navigation Wootton Wawen, at the basin. Meals at or near the bar *daily (except Tue)* during licensed hours. Omelettes, steaks, snacks etc. (Henley-in-Arden 2676).

Wilmcote.

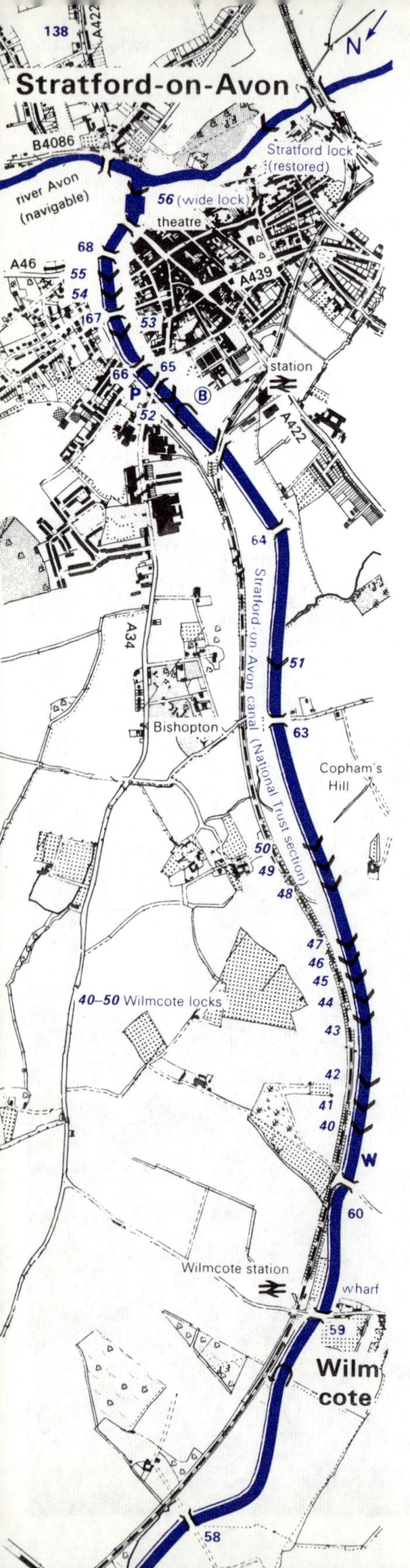

Stratford-on-Avon (National Trust section)

Stratford-on-Avon

4¼ miles

South of Wilmcote, the two long pounds from Preston Bagot are terminated by a dense flight of locks—there are eleven in the Wilmcote flight, in groups of three, five and three. They are set in pleasant open country, in which Stratford can occasionally be seen to the east. Meanwhile the countryside seems flatter, even though there are more locks to come. At bridge 64, a slight bend suddenly reveals the nether regions—gas holders and industrial works—of Stratford. Past 2 railway bridges, a grassy wharf on the right is a boatyard attached to an engineering works. At the next bridge is a winding hole, petrol station and telephone box; then the canal disappears down its own private and inaccessible corridor towards the river Avon. It drops steeply through several locks, accompanied by the little-used towpath; then the towpath disappears altogether, the canal passes through the lowest bridge since Lapworth and one suddenly emerges at the splendid great basin in the middle of the riverside parkland beside the Shakespeare Memorial Theatre. The contrast is astonishing—here is an unwalled, public and attractive basin, constantly surrounded by the famous Stratford tourists, while only a few yards away the canal is completely shut away in a world of its own, forgotten by all except boatmen. This seems all the more odd in view of the attention the canal received only a few years ago when it was restored, and ceremonially re-opened by the Queen Mother. For some reason the fine basin has no mooring rings or bollards, so boatmen have to drive stakes into the basin surrounds.

The river Avon
The wide lock at Stratford Basin leads down into the river Avon. This very pleasant waterway is the one that rises at Welford on the Leicestershire and Northamptonshire boundary near the Leicester section of the Grand Union Canal and passes through Rugby, Warwick, Stratford and Evesham before entering the river Severn at Tewkesbury. In the 17thC it was made navigable from above Stratford through to Tewkesbury, but it had a chequered history. The Avon ceased to be a through route 100 years ago, and its structures decayed steadily for 80 years. However the determined efforts starting in 1949 and led by C. D. Barwell to restore the lower Avon (from Tewkesbury up to Evesham) were successful, and by 1964 Evesham could once again be reached by boat. Since then, attention has focussed strongly on the Upper Avon. Led by the tireless David Hutchings, MBE, a team of voluntary labourers (mostly soldiers, prisoners and waterway enthusiasts) has worked constantly since 1968 on dredging and clearing, and completely rebuilding locks and weirs. The navigation is now open to within a few miles of Stratford, and by the end of 1973 boats of fairly shallow draught will be able to navigate from Stratford-on-Avon to Tewkesbury. With the Stratford and Worcester & Birmingham canals and the river Severn, the reopening of the Avon will thus restore a 110 mile circuit of splendid cruising waterways. At the moment navigation on the river at Stratford is limited to a pretty 3½-mile stretch between the village of Alveston and the already-completed top lock on the upper Avon. It is certainly worth passing through the lock from Stratford Basin down to the Avon for a cruise along the Stratford waterfront. Upstream (towards Alveston) the river becomes shallow until navigation is halted by a large fixed sluice across the river. But the whole reach at Stratford is heavily used by trip boats, small cruisers, whiffs, rowing boats and endless punts, which may be hired near Clopton bridge.

Stratford-on-Avon
Warwicks. Pop 19,000. EC Thur. MD Wed, Fri. All services. Tourism has been

established a very long time in Stratford, ever since 1789 when the first big celebrations in William Shakespeare's honour were organised by the actor David Garrick. They are now held annually on St George's Day–23rd April, which is believed to be Shakespeare's birthday. An annual Mop Fair on the 12th October reminds the visitor that Stratford was already well established as a market town long before Shakespeare's time. (The first grant for a weekly market was given by King John in 1196). Today, Stratford is well used to the constant flow of charabancs and tourists, managing successfully to retain its ancient charm unspoilt by the traditional British gaudiness that usually mars popular places like this. (Fish and chips are hard to come by in Stratford). There are wide streets of endless low, timbered buildings that house dignified hotels and antique shops; plenty of these are also private houses. On the river, hired punts and rowing boats jostle each other while people picnic on the open parkland that flanks it. The Royal Shakespeare Theatre, opened in 1932, is a splendid institution on an enviable site beside the Avon, but the aesthetic appeal of its massive 'industrial' style is limited. It was designed by Elizabeth Scott to replace an earlier theatre, destroyed by fire in 1926.

Shakespeare Birthplace Trust
This Trust was founded in 1847 to look after the 5 buildings most closely associated with Shakespeare. 4 of these are in Stratford (listed below) and the other is Mary Arden's cottage at Wilmcote (see page 136). *There is a standard opening time for these properties; any variation on this standard is shown in the individual entries. Open Apr-Oct weekdays 9.00–18.00, Sun 14.00–18.00. Nov-Mar weekdays only 9.00–16.00.*

Shakespeare's Birthplace Henley st. An early 16thC half-timbered building containing books, manuscripts and exhibits associated with Shakespeare and rooms furnished in period style. Next door is the Shakespeare centre. *Open Sun Nov-Mar, 14.00–16.30.*

Hall's Croft Old Town. A Tudor house complete with period furniture–the home of Shakespeare's daughter Susanna and her husband Dr John Hall.

New Place Chapel st. The foundations of Shakespeare's last home set in a replica of an Elizabethan garden.

Anne Hathaway's Cottage Shottery. 1 mile west of Stratford. Dating from the 15thC this fine thatched farmhouse was once the home of Anne Hathaway before she married William Shakespeare. It has a mature, typically English garden, and long queues of visitors in the summer. The cottage was badly damaged by fire in 1969, but has since been completely restored. *Open Sun Nov-Mar, 14.00–16.30.*

Holy Trinity Church attractively situated among trees overlooking the newly restored lock on the river Avon. It is mainly of the 15thC but the spire was rebuilt in 1763. Interesting misericords depict some amusing scenes, and fine monuments include one of William Shakespeare who is buried in the chancel. His tomb bears a curse against anyone who disturbs it.

Shakespeare Memorial Theatre (box office 2271). The home of the Royal Shakespeare Company, who produce Shakespeare plays to a very high standard *from April to December every year.*

Clopton Bridge
A very fine stone bridge over the Avon, originally built by Sir Hugh Clopton in about 1480-90–he later became Lord Mayor of London. The bridge is close to the canal basin. The brick bridge nearby was built in 1823 to carry a horse-drawn tramway connecting Stratford with Shipston-on-Stour. It is now a footbridge.

Information Centre Judith Shakespeare's House, 1 High st., Stratford-on-Avon (3127). Judith Shakespeare was William's younger daughter. In 1616 she married and moved into this former tavern, once called The Cage. It is a characteristic Elizabethan building of three storeys. Judith Shakespeare is buried in the graveyard of Holy Trinity church in Stratford.

BOATYARDS

Ⓑ **Western Cruisers** Western road, Stratford-on-Avon, Warwicks. (3878). Hire cruisers. Water, gas, diesel. Refuse and sewage disposal. Moorings, canal ware and books.

PUBS

There are endless pubs, hotels and restaurants in Stratford, but none actually on the canal.

X🍷 **Marianne** 3 Greenhill st, Stratford-on-Avon (3563). A fairly new French restaurant near the station. Eat bistrostyle downstairs or in more sophisticated surroundings upstairs. Seasonal variations are made to a menu of traditional French dishes–piperade, onion soup, daube provencale, coquilles St Jacques and sorbet au citron. *Lunch 12.30–14.00, dinner 19.00–23.00. (Closed Sun).* Booking essential. In the Good Food Guide since 1971.

YOUTH HOSTEL

Alveston Youth Hostel Alveston (2 miles NE of Stratford, along B4086 or up the river).

HOLIDAY FELLOWSHIP

The Fold Payton st, Stratford-on-Avon. Holiday centre on which 'Exploring Canals' holidays are based.

The river Avon in Evesham: a new lock-keeper's cottage.

Canal Books

Even when you are canal cruising you may be able to obtain the books which interest you. The British Waterways Board have a wide selection of canal books and pamphlets at Farmer's Bridge, Birmingham (The Canal Shop, 2 Kingston Row, Birmingham B1 2NU) and at Stoke Bruerne (The Waterways Museum, Stoke Bruerne, Nr. Towcester, Northamptonshire NN12 7SE). Both are on the canal. Many boatyards on the canals, especially those offering chandlery, have a selection of books for sale. Your local bookseller may only have a small stock of canal books but will order others. In addition the Inland Waterways Association (114 Regent's Park Road, London NW1 8UQ) has an extensive range, and St. John Thomas Booksellers (30 Woburn Place, London WC1 0JR) specialise in subjects such as transport.

The books below, apart from those dealing with the whole of the waterways system, are specifically concerned with the waterways of the South West (including some non-B.W.B. navigations). Publications covering canals in other parts of the country will be mentioned in the book lists of the other guides in this series.

Reference Works	**Author**	**Publisher**
Inland Waterways of Great Britain and Northern Ireland	L. A. Edwards	Imray
Inland Navigation of England & Wales: Maps of 1836	Thomas Moule	Brian Stevens, Monmouth
A General History of Inland Navigation (reprint from 1805)	J. Phillips	David & Charles
Historical Account of Navigable Rivers & Canals of Gt. Britain (reprint from 1831); also in paperback	J. Priestley	David & Charles
Bradshaw's Canals & Navigable Rivers of England & Wales (reprint from 1904); also in paperback	Henry de Salis	David & Charles
Cruising Books		
The Brecon & Abergavenny Canal (Cruising map & guide 15p)		B.W.B.
The B.C.N.: A Cruising Guide	K. D. Dunham & R. M. Manion	Inland Waterways Association
The Canal Enthusiast's Handbook	Charles Hadfield (Ed.)	David & Charles
Holiday Cruising on Inland Waterways (also in paperback)	Charles Hadfield & Michael Streat	David & Charles
Chart of the Kennet & Avon (Western half). Avonmouth–All Cannings. Chart 3 miles to 1 inch; showing all aspects of the canal.	Nicholas Hammond	Imray
The Canals Book (annual)		Link House Publications
Canal Cruising	John Hankinson	Ward Lock
Stratford-upon-Avon Canal Guide		Stratford-upon-Avon Canal Society
Map of the Stratford-upon-Avon Canal		National Trust
Historical Interest		
Industrial Archaeology of the Tamar Valley (chapter on Tavistock Canal)	Frank Booker	David & Charles
James Brindley, Engineer, 1716–1772	C. T. G. Boucher	Goose & Son
The Canal Builders	Anthony Burton	Eyre Methuen
The Dorset & Somerset Canal	Kenneth R. Clew	David & Charles
The Kennet & Avon Canal	Kenneth R. Clew	David & Charles
The Somersetshire Coal Canal & Railways	Kenneth R. Clew	David & Charles
The Wiltshire & Berkshire Canal	L. J. Dalby	Oakwood Press
Grand Western Canal		Dartington Amenity Research Trust
The Haytor Granite Tramway & Stover Canal	M. C. Evans	David & Charles
English Canals (in 3 volumes)	D. D. Gladwin & J. M. White	Oakwood Press
The Canals of South Wales & the Border	Charles Hadfield	David & Charles
British Canals (also in paperback)	Charles Hadfield	David & Charles
The Canals of South West England	Charles Hadfield	David & Charles
The Canal Age (also in paperback)	Charles Hadfield	David & Charles and Pan Books
The Canals of the West Midlands	Charles Hadfield	David & Charles
Waterways to Stratford	Charles Hadfield & John Norris	David & Charles
The Bude Canal	H. Harris & M. Ellis	David & Charles
A Tour of the Grand Junction Canal in 1819 (reprinted 1968)	J. Hassell	Cranfield & Bonfiel
The Thames & Severn Canal	H. Household	David & Charles
Development of Transportation in Modern England	W. T. Jackman	Frank Cass
The Birmingham Canal Navigations	Robert May	C. R. Smith
The Canal Duke	Hugh Malet	David & Charles
Canal and River Craft in Pictures (also in paperback)	Hugh McKnight	David & Charles
Thomas Telford	L. T. C. Rolt	Longmans
James Watt	L. T. C. Rolt	Batsford
The Severn Bore	S. W. Rowbottom	David & Charles
Lives of the Engineers (in three volumes: reprint from 1862)	Samuel Smiles	David & Charles
Waterways Heritage (paper bound)	P. Smith	Luton Museum
The Flower of Gloster (reprint from 1911)	Temple Thurston	David & Charles
River Navigation in England 1600–1750	T. S. Willan	Frank Cass
General Books		
Handbook of Water Plants	E. Bursche	Frederick Warne
Let's explore Old Waterways in Devon	A. L. Clamp	Westway Publications
The Decorative Arts of the Mariner (Small section on canals)	G. F. Cook (Ed.)	Cassell
Slow Boat through England (also in paperback)	Frederic Doerflinger	Allan Wingate
Canals in Camera (Vols. I & II)	John Gagg	Ian Allan
Canals and their Architecture	Robert Harris	Hugh Evelyn
Journeys of the Swan	John Liley	Allen & Unwin
The Canals of England	Eric de Maré	The Architectural Press

iscovering Canals (A motorist's guide)	Leon Metcalfe & John Vince	Shire Publications
/ater Highways	David Owen	Phoenix House
/ater Rallies	David Owen	Phoenix House
arrow Boat	L. T. C. Rolt	Eyre & Spottiswoode
avigable Waterways	L. T. C. Rolt	Longmans
he Inland Waterways of England	L. T. C. Rolt	Allen & Unwin
anal Fishing	Kenneth Seaman	Barrie & Jackson
oyage into England	John Seymour	David & Charles
laiden's Trip	Emma Smith	Penguin
old on a Minute	T. Wilkinson	Allen & Unwin

ooks for Young People

un on the Waterways	J. Banks & P. Hume	Penwork
/aterways Atlas of the British Isles	John Cranfield & Michael Bonfiel	Cranfield & Bonfiel
anals of the World	Charles Hadfield	Basil Blackwell
he Cow who fell in the Canal (Drawings coloured nd to colour)	Phyllis Krasilovsky	Penguin
urlew on the Cut	Beatrice Lawrence	Geoffrey Dibb
our Book of Waterways	Eric de Maré	Faber
anals in Britain	Alison Ross	Basil Blackwell
iver & Canal Transport (Approaches to nvironmental Studies Book 7)	John Vince	Blandford
he Transport Revolution (Focus on History Series aperback)	Roger Watson	Longmans
ritain's Inland Waterways	Roger Wickson	Methuen

pecialist Publications

he Facts about the Waterways (out of print)	British Waterways Board	HMSO 1965
eisure and the Waterways (out of print)	British Waterways Board	HMSO 1967
he Last Ten Years	British Waterways Board	B.W.B. 1973
ritish Waterways Board: Annual Report and ccounts	British Waterways Board	HMSO (yearly)
irmingham Canal Navigations (a report on the emainder' waterways)	British Waterways Board	B.W.B. 1971
irmingham and its Regional Setting: a scientific urvey (chapter on canals)	M. J. Wise	S.R. Publishers, Wakefield
recon Beacons: National Park Guide No. 5	Ed. Margaret Davies, M.A., Ph.D.	HMSO 1967

rdnance Survey Maps

he following 1 inch Ordnance Survey maps refer to the canals covered by this book:

19	Stafford	Birmingham Canal Navigations, Staffordshire & Worcestershire Canal
20	Burton upon Trent	Birmingham Canal Navigations
30	Kidderminster	Severn Navigation, Staffordshire & Worcestershire and Worcester & Birmingham Canals
31	Birmingham	Birmingham Canal Navigations, Stourbridge, Worcester & Birmingham, Staffordshire & Worcestershire and Stratford on Avon Canals
41	Brecon	Monmouthshire & Brecon Canal
42	Hereford	Monmouthshire & Brecon Canal
43	Gloucester & Malvern	Severn Navigation and Gloucester & Sharpness Canal
55	Bristol & Newport	Brecon & Abergavenny Canal
56	Bristol & Stroud	Kennet & Avon and Gloucester & Sharpness Canal
58	Oxford & Newbury	Kennet & Avon Canal
59	The Chilterns	Kennet & Avon Canal
65	Weston super Mare	Bridgwater & Taunton Canal
66	Frome	Kennet & Avon Canal
67	Salisbury	Kennet & Avon Canal
77	Taunton & Lyme Regis	Bridgwater & Taunton Canal

he Ordnance Survey also produce a complete set of $2\frac{1}{2}$ inch maps, flat or folded, which are ideal for rge-scale representation of the canals and show every lock, bridge, etc. (The maps in this guide are ased on these sheets.)

n addition, David & Charles publish a reprint of the First Edition of the Ordnance Survey, dating from 860 to 1870. 97 separate sheets cover England and Wales.

3oat Clubs

he following is a list of boat clubs on waterways covered by this guide book, lassified by the navigation on which each club has its main mooring site.

irmingham Canal Navigations
irmingham Rowing Club
abrina Cruising Club
ongwood Boat Club

/orcester & Birmingham Canal
ing's Norton Motor Cruising Club

evern Navigation
iloucester Yacht Club
iloucester Rowing Club
Valls Sailing Club
/orcester Rowing Club
ewkesbury Cruising & Sailing Club
hornbury Sailing Club
evern Motor Yacht Club

Kennet & Avon Canal
Burghfield Island Boat Club
Bristol Avon Sailing Club
Cabot Cruising Club
Newbury Canoe Club

Bridgwater & Taunton Canal
Creech St Michael Boating Club

Monmouthshire & Brecon Canal
Govilon Boat Club

Stratford upon Avon Canal
Earlswood Motor Yacht Club

Staffordshire & Worcestershire Canal
Stafford Boat Club
Stourport Boat Club
Stourport Yacht Club

An up to date list of the clubs shown may be obtained from the British Waterways Board at Melbury House, Melbury Terrace, London NW1 6JX. This list gives the name and address of each club's current Honorary Secretary.

Prospective canoeists, campers and walkers may find the following addresses useful:

British Canoe Union
26 Park Crescent
London W1

Canoe Camping Club
11 Grosvenor Place
London SW1

Ramblers' Association
1 Crawford Mews
York street
London W1

Youth Hostels Association
29 John Adam street
London WC2

Canal Societies

The Inland Waterways Association is 'the' national society of canal enthusiasts. It is the oldest, biggest and most influential of the canal societies and has permanent offices and staff. It has over 7,000 members and publishes a regular Waterways 'Bulletin'. There are subsidiary branches of the IWA in South West England and South Wales: their addresses can be obtained from the General Office at 114 Regent's Park Road, London NW1 8UQ.

Local canal societies and other interested groups in the South West include:

Birmingham Canal Navigations Society
Bridgwater YMCA Centre Canal Club
Dudley Canal Trust
Exeter Maritime Museum
Grand Western Canal Preservation Committee
Hockley Port Youth Project, Birmingham
Kennet & Avon Canal Trust
Lower Avon Navigation Trust
Newport (Mon.) Canal Preservation Society
Railway & Canal Historical Society
Risca Magor & St. Mellons Canal Preservation Society
Somerset Inland Waterways Society
Southampton Canal Society
Staffordshire & Worcestershire Canal Society
Stratford upon Avon Canal Society
Stroudwater Canal Society
Worcester & Birmingham Canal Society

An up-to-date list of canal societies throughout the country may be obtained from the British Waterways Board at Melbury House, Melbury Terrace, London NW1 6JX. This list gives the name and address of each society's current Honorary Secretary.

BWB Offices and Yards

Headquarters

Melbury House, Melbury Terrace, London NW1 6JX (01 262 6711). General and official enquiries and correspondence.

Willow Grange, Church road, Watford, Herts. (Watford 26422). Pleasure craft licences and registration, mooring permits and angling arrangements.

Gloucester Area Engineer

Dock Office, Gloucester (25524). Responsible for maintaining the waterways listed below, via the BWB Section Inspector or Foreman at the maintenance yards on those canals. The numbers in brackets refer to the pages in this book on which the yards occur.

Severn Navigation Diglis (81)
Gloucester & Sharpness Canal Sharpness (70)
Brecon & Abergavenny Canal Gilwern (55)
Bridgwater & Taunton Durston (64)
Kennet & Avon Canal Devizes (38)
Padworth (26)

Birmingham Area Engineer

Reservoir House, Icknield Port road, Birmingham (021-454 7091). Responsible for maintaining the waterways listed below.

Birmingham Canal Navigations Birmingham (122)
Sneyd (122)
Worcester & Birmingham Canal Tardebigge (105)
Staffordshire & Worcestershire Canal Stourport (90)
Stourbridge Canal
Stratford upon Avon Canal

Amenity Services Supervisor (North West)

Nantwich Basin, Chester road, Nantwich, Ches. (65122). General enquiries about cruising in the north west, and permits for fishing from boats. Craft licences and mooring permits can be obtained by calling at this office: postal application should be made direct to the Craft Licensing Officer at Watford (see above).

Amenity Services Superintendent (Midlands)

Canal Shop & Information Centre, 2 Kingston Row, Birmingham B1 2NU (021-236 4844). General enquiries about cruising in the Midlands and about the hire of canal cruisers from Nantwich and Hillmorton. Craft licences and mooring permits can be obtained by calling at this office; postal applications should be made direct to the Craft Licensing Officer at Watford.

Index